ATHENIAN CALENDARS AND EKKLESIAS

ΑΡΧΑΙΑ ΕΛΛΑΣ

MONOGRAPHS ON ANCIENT GREEK HISTORY AND ARCHAEOLOGY

VOLUME 8

ATHENIAN CALENDARS AND EKKLESIAS

W. KENDRICK PRITCHETT

EMERITUS PROFESSOR OF GREEK IN THE
UNIVERSITY OF CALIFORNIA, BERKELEY

J.C. GIEBEN, PUBLISHER
AMSTERDAM 2001

 / Printed in The Netherlands / ISBN 90 5063 258 0

IN MEMORIAM

ELIZABETH DOW PRITCHETT

1.2.1911 – 3.13.2000

CONTENTS

ACKNOWLEDGEMENTS

This book could not have been published without the assistance of Barbara Saylor Rodgers, chairperson of the Classics Department at the University of Vermont, who has typed semi-final and final drafts and then formatted the text onto a floppy disk. To her I owe a debt which goes beyond the ability of words to acknowledge. I am indebted to Ronald S. Stroud and Mary C. Richardson who responded to my requests in publishable form for readings on stones in Athens. As with *SEG*, J. C. "Hans" Gieben continues to be the ideal publisher for epigraphical material.

INTRODUCTION

This work purports to be an evaluation of the criticisms which were brought against *Calendars of Athens* published with the collaboration of Otto Neugebauer in 1947. Separate features of our theories might be treated in articles, but the subject of the nature of the Athenian festival and prytany calendars has so many overlapping features, viz., the epoch of the prytany calendar, the epoch of the day, irregularities in the intercalating and subtraction of days in the festival calendar, observation of the lunar movement, which, when treated in isolation, has led to dissembling results. The restored texts of slivers of fragments of prescripts only two or three stoichoi wide has been evoked to support theories about the omitted day in hollow months, although the claim for Day 22 lacks any support in Greek or world astronomy. The shadowy figures of Solon and Meton have been created as the great innovators of Athenian time-reckoning, the latter even for the prytany calendar. In the most recent volume of Agora inscriptions (1997), each archon is correlated with what the author calls the Metonic cycle, although this so-called cycle is not used to predict the distribution of ordinary and intercalary years. Extracts have been taken from Geminos, who never mentions Athens, completely out of context. The zodiacal parapegma of 365 days appended to his treatise names Meton only once. Rules are invoked for one archon and one stone-cutter which are denied to others. All these and related subjects require elucidation in one central treatise.

This book is concerned only with the principles of Athenian time-reckoning, the nature of their festival and prytany calendars. In *Calendars of Athens*, we attempted to collect all calendar equations and classify them according to ordinary and intercalary years. But the subject of the character of a year is far more elusive than evoking a theory of least irregularity. In *The Choiseul Marble* 80-81, 87, we observed that one scholar was led to alter the designations of the characters of years (whether ordinary or intercalary, whether certain or uncertain) in thirty-six cases in publishing an archon-table as compared to an earlier one by himself. The sequence of archons lies outside the scope of this work. Only selected calendar equations which relate to our problems are treated here. The appearance of A. G. Woodhead's *The Athenian Agora*, Volume XVI: Inscriptions (1997) and several isolated but very incomplete articles on the Athenian calendars are justification for this study.

The names of each Attic month with a Roman numeral are as follows:

I	Hekatombaion	V	Maimakterion	IX	Elaphebolion
II	Metageitnion	VI	Posideon	X	Mounichion
III	Boedromion	VII	Gamelion	XI	Thargelion
IV	Pyanopsion	VIII	Anthesterion	XII	Skirophorion

The various terms for days in the festival calendar as they occur in inscriptions and literary texts, including some scholia, are given below. Some specific days are hypothetical; when ἐσιόντος and ἀπιόντος occur with one ordinal in the second century, I assume that they would be used with other ordinals of the decade. The term δεκάτη φθίνοντος is not found in inscriptions, but δεκάτη ἀπιόντος occurs in the apocryphal decree in Demosthenes 18 *Crown* 37. Alphabetical numerals are found in manuscripts such as the *Ath. Pol.* and scholia. Not all days are attested. For the best collection of days in literary sources, see A. Mommsen, *Chronologie* (Leipzig 1883) 80-123.

1.	νουμηνία	α´
2.	δευτέρα ἱσταμένου, ἐσιόντος	β´
3.	τρίτη ἱσταμένου, ἐσιόντος	γ´
4.	τετρὰς ἱσταμένου, ἐσιόντος	δ´
5.	πέμπτη ἱσταμένου, ἐσιόντος	ε´
6.	ἕκτη ἱσταμένου, ἐσιόντος	ς´
7.	ἑβδόμη ἱσταμένου, ἐσιόντος	ζ´
8.	ὀγδόη ἱσταμένου, ἐσιόντος	η´
9.	ἐνάτη ἱσταμένου, ἐσιόντος	θ´
10.	δεκάτη ἱσταμένου, ἐσιόντος	ι´
11.	ἑνδεκάτη	ια´
12.	δωδεκάτη	ιβ´
13.	τρίτη ἐπὶ δέκα	ιγ´
14.	τετρὰς ἐπὶ δέκα	ιδ´
15.	πέμπτη ἐπὶ δέκα, διχομηνία (*IG* I^3 6)	ιε´
16.	ἕκτη ἐπὶ δέκα	ις´
17.	ἑβδόμη ἐπὶ δέκα	ιζ´
18.	ὀγδόη ἐπὶ δέκα	ιη´
19.	ἐνάτη ἐπὶ δέκα	ιθ´
20.	εἰκάς, εἰκοστή, δεκάτη προτέρα, δεκάτη προτεραία	κ´
21.	δεκάτη φθίνοντος, δεκάτη ὑστέρα, δεκάτη ἀπιόντος	κα´
22.	ἐνάτη φθίνοντος, μετ' εἰκάδας, ἀπιόντος	κθ´
23.	ὀγδόη φθίνοντος, μετ' εἰκάδας, ἀπιόντος	κη´
24.	ἑβδόμη φθίνοντος, μετ' εἰκάδας, ἀπιόντος	κζ´
25.	ἕκτη φθίνοντος, μετ' εἰκάδας, ἀπιόντος	κς´
26.	πέμπτη φθίνοντος, μετ' εἰκάδας, ἀπιόντος	κε´
27.	τετρὰς φθίνοντος, μετ' εἰκάδας, ἀπιόντος	κδ´
28.	τρίτη φθίνοντος, μετ' εἰκάδας, ἀπιόντος	κγ´
29.	(full month) δευτέρα φθίνοντος, μετ' εἰκάδας, ἀπιόντος	κβ´
30.	ἕνη καὶ νέα, Δημητριάς	λ´

I

EPIGRAPHICAL EVIDENCE FOR IRREGULARITIES IN FESTIVAL CALENDAR

One element of the interlocking evidence about Athenian time-reckoning which appropriately precedes a collection of the literary evidence about the calendar is the epigraphical evidence for tampering with the festival calendar. Precedence is here given to this feature of the festival calendar because this evidence is preliminary to any analysis of its character, determined largely by restorations, and although the evidence is limited, we must formulate rules on the assumption that what one archon (or civil authorities) might do with his calendar, others might. In *The Calendars of Athens* (1947), Neugebauer and I inherited a theory that the year was made up of months alternating between full and hollow with leap days. Our collaboration began when Neugebauer observed that κατὰ θεόν in some epigraphical texts was synonymous with κατὰ Σελήνην and denoted a lunar calendar, a calendar which was operative if the archon did not intercalate or suppress dates, an untampered festival calendar. My purpose here is to bring up-to-date epigraphical evidence about dates which deviated from this calendar, including those designated as κατ' ἄρχοντα or ἐμβόλιμος.

Table 1 presents examples of (1) kat' archonta, (2) kata theon, and (3) prytany dates found in Attic inscriptions. Included are three examples of Meritt based on the fact that a long lacuna occurs in the text, for which triple dates are a possibility. They are at best exempli gratia. The dates of many second-century archons are debated; those given do not represent any personal research. C. Habicht, *Athens from Alexander to Antony* (1997) p. v, writes,

"The chronology of archons of the period will result in numerous changes of accepted archon-years." See also M. J. Osborne's review of G. Woodhead's *Athenian Agora* XVI (1997) in *AJA* 103 (1999) 710–711.

I make the observation here, which will be discussed when we come to the prytany calendar, that in two of the ordinary years (Demetrios and Euergetes) the prytany and kata theon dates differ by a day. This proves, as was observed in *Calendars of Athens* p. 78, that the two calendars were not predetermined to march pari passu. The difference accords with our theory that one was lunar, the other predetermined, and I believe according to the Aristotelian pattern. Woodhead, *Agora* XVI p. 413, writes of the Euergetes equation, "The prytany calendar appears to be out of complete accord with the calendar κατὰ θεόν, but there was evidently, in this year, no exact correspondence between 'natural' months and prytanies." The word natural in quotation marks is new in the calendric literature, but suggests to me lunar.

Difference between dates κατ' ἄρχοντα and κατὰ θεόν in Table 1.

1 day in Nos. 3, 20, 21, 22
2 days in Nos. 10, 13
3 days in No. 14
4 days in No. 9
7 days in No. 2
11 days in No. 17
20 days in No. 19
28 days in No. 5
Month in No. 6

TABLE 1
TRIPLE DATES, ALLEGED AND TRUE

No.	Text	Year	Archon	Type of Year	κατ᾽ ἄρχοντα Date	κατὰ θεόν Date	Prytany Date
1	*IG* II^2 861	fin. III	?	Space = Triple date (Koehler) or embolimoi days (Meritt, *Year* 66)			
2	*Hesp*. 33.183[a]	init. II	?	I	XI 11	XI 18	XI 23
3	*Chron*. 112[b]	193/2	Phanarchides		[II] 2[2]	[2] 3	II 3
4	*Hesp*. 5.428[c]	191/0 179/8?	?		Omitted	k.th.	? 9
5	*IG* II^2 946	190/89	Achaios	O	VIII [2] 9	IX 27 [k.th.]	[IX 27]
6	*IG* II^2 947	190/89	Achaios	O	X 12	XI [11/12 k.th.]	XI 12

[a] In *Agora* XV no. 207, Meritt reassigned this inscription to archon Alexis (173/2). In *Agora* XVI, it is not republished by Woodhead under this archon or anywhere else that I can find.

[b] B. D. Meritt, *Hesperia* 32 (1963) 16-17, proposes [Hekatombaion (I) ἕκτηι μετ᾽] εἰκάδας ὀγδ[όει ἐμβολίμωι], which seems preferable.

[c] Demotic of secretary is preserved. All that is certain is ον presumably of κατὰ θεόν. Tracy, *Attic Letter-cutters* 84, dates mason to 203/2-184/3. See Meritt, *Mnem*. 26 (1976) 162.

No.	Text	Year	Archon	Type of Year	κατ' ἄρχοντα Date	κατὰ θεόν Date	Prytany Date
7	*SEG* 16 no. 92[d]	173/2	[Alexis]	I	Omitted	[X] 11 [k.th.]	[X 18]
8	*SEG* 16 no. 93	173/2	Alexis	I	Omitted	X 11 [k.th.]	[X 18]
9	*Chron.* 121[e]	170/69	Demetrios	O	IX 23	27 [k.th.]	[IX] 28
10	*Hesp.* 26.72	164/3	Euergetes	O	IX 19	IX 21 k.th.	IX 22
11	*SEG* 34.95.2	161/0	Aristolaos	O	[III ?]	16 k.th.	III 16
12	*SEG* 34.95.26	161/0	Aristolaos	O	2-10	? k.th.	---
13	*SEG* 34.95.44	161/0	Aristolaos	O	VI 22	VI 24 k.th.	----
14	*IG* II² 979	155/4	Mnesitheos	O	[?] 21	2[4] k.th.	? [2]4
15	*Hesp.* 10.61[f]	155/4	Mnesitheos	O	Two sure letters preserved. Restorations of BDM		
16	*IG* II² 1001[g]	med s.II	?	?	(?) k.a.	VII 10	?
17	*Hesp.* 11.295	152/1 or 148/7	Lysiades	O	[IV] 5 k.a.	IV 16 k.th.	- - -

[d] Four letters preserved; otherwise restoration.

[e] See *Agora* XV no. 171; Woodhead, *Agora* XVI.405; Habicht, *Hesperia* 57 (1988) 245.

[f] Woodhead, *Agora* XVI 421, repeats Meritt's equation: 4 (kat' archonta) = 6 kata theon = prytany 6th day of an unknown month and prytany. Prime example of game of restoration on basis of two letters in non-stoichedon, almost illegible inscription.

[g] M. Walbank, *BSA* 85 (1990) 441-442 (= *SEG* 40 no. 69) writes that this fragment was misread, being the same as *IG* II² 321, dated before 336/5, which rules out the calendric restoration of *IG* II² 1001, perpetuated in all of our earlier lists.

No.	Text	Year	Archon	Type of Year	κατ' ἄρχοντα Date	κατὰ θεόν Date	Prytany Date
18	*Hesp.* 34 .90[h]	152/1 or 148/7	Lysiades		[IV 8^3]	[IV 22]	[IV 22]
19	*IG* II^2 967	145/4	Metrophanes	O	IX 22 k.a.	X 12 k.t.	X 12
20	*Hesp.* 15.207	127/6	Theodorides	I	III 27 (k.a.)	[28/9] k.th.	[III] 24
21	*IG* II^2 1004	122/1	Nikodemos	O	III 8^2 k.a.	9 k.th.	III 9
22	*IG* II^2 1006A	122/1	Nikodemos	O	III 8^2 k.a.	9 k.t.	III 9
23	*Hesp.* 10.61[i]	122/1	Nikodemos	O	All restorations by BDM except for name of month V		
24	*Agora* XV no. 261II	95/4	Theodotos		? k.a.	V [8]	V 8
25	*Agora* XV no.261I	95/4	Theodotos		[k.a.]	[V 9] k.th.	V 9

[h] Six letters. Entirely restoration.

[i] For the lettering with photograph, see Tracy, *Attic Letter-Cutters* 192-193.

In the kat' archonta and kata theon dates, all months are represented except Hekatombaion (I) and Skirophorion (XII).[1]

A second type of epigraphical evidence for irregularities in the festival calendar is month-dates in Athenian prescripts designated as ἐμβόλιμος, "intercalated".[2] At the time of the *Calendars of Athens* (p. 22), the highest date so designated was δευτέρα ἐμβόλιμος. Dinsmoor, and Meritt early in his career, took the term, although in the dative case, to apply to the month. More recently, tertarte embolimos has turned up on the stone, and Meritt's partial restoration of ogdoe embolimos seems most probable. Table 2 presents examples including three which should be termed only exempli gratia in long lacunae.

TABLE 2

TABLE OF EMBOLIMOI DAYS: TRUE AND ALLEGED

	REFERENCE	DATE
Embolimos		
Hekatombaion² 25	Dow, *Prytaneis* No. 29	228
Boedromion 8	*IG* II² 1006A	122
Elaphebolion 29/30 (2)	*IG* II² 358, 791	333, 221
Mounichion 29/30	*IG* II² 471	305
Deutera embolimos		
Metageitnion [1]6	*Hesp.* 23 236	214
Metageitnion 22	*IG* II² 838	226
Gamelion 23	*IG* II² 458	306

[1] A. Mommsen, *Feste der Stadt Athen* (1898) 171, observed that Boedromion with the celebration for Marathon, the Genesia, and the Eleusinian Mysteries, was a month full of festivals.

[2] See Dio C. 48.33.4; Hiller, *Inschriften von Priene* no. 105 line 76 (9 B.C.).

	REFERENCE	DATE
Trite embolimos		
[Pyanopsion 8]	*Hesp.* 34 90 (entirely restoration of Meritt)	148
Tetarte embolimos		
Hekatombaion 2[2+]	*Year* 166	III s.
Anthesterion 11	*AE* 1968 115 (entirely restoration of Meritt)	306
Elaphebolion 9	*Hesp.* 23 299	270
Hebdome embolimos		
Anthesterion 11	*AE* 1968 113 (restoration of Meritt)	306
Ogdoe embolimos		
Hekatombaion 25	*Hesp.* 32 17	193

MONTHS. In this table, we find embolimoi days in six different months: Hekatombaion (I), Metageitnion (II), Boedromion (III), Gamelion (VII), Elaphebolion (IX), Mounichion (X), with Meritt adding Pyanopsion (IV) and Anthesterion (VIII) by dubious restorations. The month not represented in either table is Skirophorion (XII).[3] This was the last month of the year with the fewest festivals and, as we shall see, a candidate for exairesimoi, rather than embolimoi, days.

Irregularities were not limited to days. The same is true of months. I list below examples where an intercalary month is named in Athenian inscriptions before 100 B.C. These examples are restricted to instances where the critical word for intercalation, or part of it, is found on the stone.[4]

[3]Jon D. Mikalson, *The Sacred and Civil Calendar of the Athenian Year* (1975) 182, finds that only four preserved decrees of the ekklesia are dated in Maimakterion. Skirophorion provides the largest number, especially for the last day, more than twice as many as for any other day of the year, a day when the boule and ekklesia cleared the slate.

[4]J. H. Kent, *Hesperia* 16 (1947) 224, publishes an inscription from Delos which has a second Posideon in the archonship of Anthesterios (II

TABLE 3

MONTH	SOURCE	DATE
Posideon ἐμβόλιμος (2)	*IG* II^2 785; *Year* p. 195	Charikles: II cent.; Hippias: II cent.
Posideon ὕστερος (2)	*IG* II^2 381+382; 683	Neaichmos: IV cent.; Hieron: III cent.
Posideon δεύτερος	*IG* II^2 917	II cent.
Intercalated Hekatombaion	*IG* I^3 78	V cent.
Hekatombaion ὕστερος	Dow, *Prytaneis*, No. 29	Leochares: III cent.
Metageitnion ἐμβόλιμος	*Hesperia* XXVI (1957) 34	Alexis: II cent.
Gamelion ὕστερος	*IG* II^2 1487.53-54	Anaxikrates: IV cent.
Anthesterion ἐμβόλιμος	*IG* II^2 844	Archelaus: III cent.

Of the evidence from ten different intercalary years, five, or 50 per cent., exhibit a second Posideon. More importantly, five different months were intercalated The position of the intercalary months was after Months I, II, VI, VII, and VIII. One expects schematic calendars to follow a rigid pattern of intercalation. Such rigidity was not the case at Athens.[5]

One of the most informative texts about the duration of irregularities in a single year is an Agora inscription giving an inventory of dedications in a sanctuary, partly published by Meritt for which S. Tracy made several important corrections in the readings. The text records three κατὰ θεόν days in a period from

---

cent.). The restoration in *IG* II^2 1290 also looks secure for a second Posideon.

[5]Long ago, Kirchner, *Sitz. d. Berl. Akad.* 1910 982 ff., suggested that irregular intercalation of months was not uncommon.

Boedromion (III) 16 to Posideon (VI) 24. The text is that of *SEG* 34.95 (161/0).[6]

SEG 34.95

Line 2: [ἐπὶ 'Αριστόλα ἄρχοντος ἐπὶ τῆς - 4 or 5 -]ίδος τρίτης πρυτανείας ... [Βοηδρομιῶνος - - - ca. 14 - - - κατὰ] θεὸν δὲ ἕκτει ἐπὶ δέκα ἕκτει κα[ὶ δεκάτει τῆς πρυτανείας·

Line 26: ἐπὶ 'Αριστόλα ἄρχο[ντος - - - ca. 15 - - -ε]ι ἱσταμένου, κατὰ θ̣[εὸν δὲ - - - - - ca. 33 - - - - -

Line 44: [- - ca. 9 - - Ποσιδ]εῶνος ε[- -ca. 5 - - μετ' εἰκά]δας, κατὰ θ̲[ε]ὸν δὲ ἑβδόμει μ̣[ετ' εἰκάδας - - - - ca. 26 - - - -]

Thus, in a period of approximately 97 days (14 + 59 + 24) as measured by the κατὰ θεόν calendar there was considerable retardation of dates in the festival calendar. We have a choice of assuming that the suppression was continued over the entire period, which seems likely, or that the archon suppressed dates on three separate occasions and then brought the kat' archonta and kata theon calendars into alignment on each occasion.[7]

Another example of lengthy retardation occurs in Nos. 5 and 6 of the Table of triple dates, both being of the same year. On the 263rd day (Prytany IX 27) of the year, the festival calendar was running 28 days behind the kata theon (VIII 29 = IX 27); 44 days later, on the 307th day, the festival calendar was 30 days behind (X 12 = XI 12). The archon had not only not corrected the

[6]Cf. *SEG* 39.139; 41.60.

[7]Tracy, *Hesperia* 53 (1984) 376, suggests line 44 gives the date the inventory was completed. He believes the inventory was that of a sanctuary of Aphrodite, as does Tréheux, *REG* 103 (1989) p. 418. Oddly, Woodhead, *Agora* XVI (1997) 415–421, republishes the text without any recognition of the duration of the irregularities in the festival calendar.

irregularity over this period, but he had introduced two additional days.

In the table of triple dates, init. s. II – 95 B.C., there are ten sure years which give complete equations. In the embolimoi table, the spread is 333 – 122 B.C. From the paucity of evidence, it would be foolish to argue that the archon had to get his calendar in order by Thargelion (XI) ultimo. In the table of triple dates, there are two with Thargelion (XI) dates in the kata theon calendar. No. 2 (*Hesperia* 33 [1964] 183) shows that the festival calendar kat' archonta was seven days behind and the retardation may have continued. The equation is Thargelion 11 k.a. = Thargelion 18 k.t. With 11/12 days left in the month the festivals of the Bendideia (19th), Plynteria (25th), and Kallynteria (?) still had to be performed. The Bendideia had a procession, a torch-race on horseback, and a παννυχίς. At the Plynteria, the ancient image of Athena Polias was undressed, washed, and probably taken in procession to the sea; all temples were closed. It is reasonable to assume that the Bendideia and Plynteria would be separated by the customary six days to allow for preparations. There is a strong probability in this example that the suppression of dates carried over into Skirophorion.

The second example of a kata theon date in Thargelion (No. 6) gives the equation: Mounichion (X) 12 = Thargelion (XI) 11/12 kata theon = Prytany XI 12. With 47 days to go, 30 were to be exairesimoi. This means that many dates in Skirophorion were suppressed. This last month of the year had the fewest attested festivals. Jon Mikalson, *Sacred and Civil Calendar* 167, suggests the Arrephoria fell on Skirophorion 3; the *Ety. Magnum* dates it in this month. The Diisoteria, of uncertain date (Deubner), was a procession through the Peiraieus to a large shrine; see R. A. Parker, *Athenian Religion* (1996) 240. The Dipolieia, for which Meritt, *Hesperia* 37 [1968] 267–268, published a decree regulating the festival (350–330 B.C.), fell on the 14th (Mikalson), but

Parker (p. 270) suggests that this festival disappeared about 300 B.C. The number of exairesimoi days at the end of the year is a relative matter. As late as the middle of the tenth month, the kat' archonta calendar was twenty days behind the kata theon (No. 19 in the Table). If years began and ended coterminously, the conclusion is inescapable that many of these 30 dates in No. 6 were suppressed in Skirophorion.

Working on the assumption that what one archon could do with his year another might, the following rules for manipulation of the festival calendar based on the above epigraphical evidence may be offered:

1. Any month might be the intercalary one.

2. Any day might be retarded up to eight days.

3. Any festival year might be retarded by 28 days by the time of the ninth month.

4. A. The retardation might be continued over a period of 97 days

or

B. In a span of 97 days retardation might occur on 3 separate occasions.

5. Of the last 45 dates in a year, 30 might be exairesimoi.

6. Retardation before the ninth lunar month was not corrected until after the beginning of the eleventh month.

Parenthetically, these rules are based on the assumption, which I, following all scholars before Meritt, believe valid, that after the end of the fifth century Hekatombaion (I) 1 = Prytany I 1. As will be documented later, Meritt, for certain selected years, denies this equation, has the prytany year with different terminal limits than the festival one, and has two archons in office at the same time.

B. D. Meritt throughout his studies and A. G. Woodhead as recently as 1997 do not agree with the interpretation of calendric irregularities as given here. Rather, Meritt invoked two rules and has been followed by Woodhead:

1. There was immediate compensation, apparently before the end of the month, for any dislocation of the calendar.

2. More importantly, there were no embolimoi or exairesimoi days in Skirophorion.

As to Rule 1, Meritt wrote in *Athenian Year*:

> Page 104: When the occasion passed, one expects the calendar to revert to normal, and I propose that this was managed before the end of the month.
>
> Page 103: Irregularities were apparently introduced, if at all, for some special whim or purpose, and—the purpose served—the irregularity was erased without long delay.

In *TAPA* 95 (1964) 246-247, he wrote:

> It is a matter of common sense that even without a "law" the archon will have at his early pleasure brought even the festival calendar back into harmony with the moon when the occasion for the dislocation was past and done.[8]

[8]At the time Meritt believed that the kata theon calendar was the "Metonic" one of Geminos 8.55–56 with all months of 30 days and every 64th day omitted. Therefore, the statement "back into harmony with the moon" is contradictory.

Meritt's practice did not always follow the rules which he laid down for others. In *Agora* XV p. 164, Meritt/Traill wrote concerning Pyanopsion (IV) 22. 178/7, "It is evident that the festival calendar has been retarded by eight days. Four of these days had been compensated before Maimakterion 9." Again, on p. 114:

> Pritchett's original restoration for this year left an initial irregularity in the civil calendar of extra days intercalated early in the year uncorrected for a span of more than six months (Pritchett and Neugebauer, *Calendars*, pp. 73–74; cf. B. D. Meritt, *The Athenian Year*, pp. 153–154). This span can be reduced to about six weeks by restoring lines 1–3 as indicated above, assuming that of the three days added to the civil calendar before Prytany III 27 one had been eliminated before Prytany IV 32 and the other two shortly thereafter.

Actually, I proposed only that there was irregularity at the early part of the year and that in the later part the festival calendar was two days behind the prytany one. There might be more than one occasion in a year for tampering with the calendar, and I left the matter open. The interesting feature is that Meritt has his calendar uncorrected for six weeks. Woodhead (*Agora* XVI [1997] 195), following Meritt, writes:

> **124** indicates that two days had been intercalated into it later than Maimakterion 12, the date of E.M. 13221, and before Posideon 29. The effect of this intercalation continued into the month Gamelion, as is shown by *IG* II^2, 499, but compensation had been made for it before *IG* II^2, 500 was passed late in Anthesterion.

In *Hesperia* 38 (1967) 111, Meritt wrote, "Two days had been intercalated late in Hekatombaion to postpone the Panathenaia and that they had not as yet been eliminated before Metageitnion 9."

As to Rule 2, Meritt wrote in *Athenian Year*:

> Page 103: I have nowhere found any evidence that extra days were ever dropped from Skirophorion.
> Page 208: In the record so far as preserved, in all centuries, fifth, fourth, third, and second, there is no example of extra days added irregularly to the festival calendar or of compensating subtraction in the month of Skirophorion, which ended the year.
> Pages 208, 209: So far as we have evidence, every known date in Skirophorion was not only κατ' ἄρχοντα but κατὰ θεόν as well. ... What I have done is to go on the assumption that the dates given are *honest* dates (italics mine).
> Page 211: We know that it is a good equation because there is no evidence anywhere for any but good equations in the festival calendar at Athens during Skirophorion or the early days of Hekatombaion.
> Page 212: In view of the fact that such dates in late Skirophorion and early Hekatombaion were never wrong, I find it astonishing that one has come to deny the *prima facie* evidence of these texts.

In *TAPA* 95 (1964) 245, he wrote: "I have said that 'so far as we have evidence' there is no example of tampering with the festival calendar in the month of Skirophorion." A. G. Woodhead in *Agora* XVI (1997) 149, wrote: "Compensation for earlier retardation was never, on the evidence available, made in Skirophorion." These quotations are but samples.

The phrase "the evidence available" must include the table of triple dates with its limited number of examples of κατ' ἄρχοντα, κατὰ θεόν and prytany dates. Neither Meritt nor Woodhead offers any summarizing table giving available evidence. Obviously such evidence includes their interpretation of these examples. Incidentally, any argument ex silentio comes with poor grace concerning the Athenian calendar. Meritt maintained the thesis that in selected examples the Prytany and Festival years were not coterminous, that Prytany I 1 did not equal Hekatombaion (I) 1 and that Hekatombaion 1 kat' archonta did not equal Hekatombaion 1 kata theon. Sample quotations are:

> *Historia* 26 (1977) 165: The final month Skirophorion of the archonship of Nikosthenes became the first month of the year of Achaios.
> *Arch. Eph.* 1968 94: The calendar κατὰ θεόν, which began earlier by a month than the calendar κατ' ἄρχοντα.
> *TAPA* 95 (1964) 244: This overlapping of festival and conciliar years was by no means new in Athenian calendar history; for such maladjustment was normal in the fifth century.
> *Historia* 26 (1977) 165: The year ... had therefore only twelve months (Skirophorion 1 to Thargelion *ultimo*).

By having the festival calendar end on Thargelion ultimo, Meritt overlooks the fact that the festivals of Skirophorion would be omitted. His second year receives an extra month, but the first month is Hekatombaion. The table of triple dates shows the manipulation of days, not months, in coterminous years. No. 6 is firm evidence that days were omitted in Skirophorion. Moreover, if there was no calendar reform c. 407/6, but all Hekatombaion 1's might equal any prytany date, we rethink all calendar equations. Codifying Meritt's practice for other archonships, chaos reigns. Neither Meritt, master of restorations, nor Woodhead offers a specimen of what the text of a decree passed in their months of dual archonship would be. All of this calendric nonsense is put forth to avoid the obvious, that days might be omitted in Skirophorion (XII). In a year studied in *Hesperia* 38 (1969) 437, Meritt posits that the prytany year began as intercalary, but the festival year as ordinary. I shall return to the subject in treating Meritt's calendar of the leap day.

Why the strange insistence on a sacrosanct Skirophorion? With Meritt's calendric theories, it is instructive to attempt to find their origin. We shall document in our opening section on the omitted day that the theory of Day 22 was put forth in *Hesperia* IV on erroneous interpretations, as Meritt later conceded, although he continued to advocate the theory without a backwards glance. In brief, the Skirophorion theory originated in Chapter 10 of *Athenian Year* (1961) and has to do with fifth-century chronology. In his *Athenian Calendar of the Fifth Century* (1924), Meritt offered two equations, which he acknowledged were the cornerstones of his chronology:[9]

423/2. [Skirophorion] 23 = Prytany X 20
412/1. Skirophorion 14 = Prytany I 1

Assuming without argumentation that the Athenians had a rigid calendar of alternating full and hollow months in which he could determine the character of all the months if he knew that of one (i.e. Skirophorion 422), he postulated a series of ordinary and intercalary years throughout an interval of 10 years. More importantly, he claimed the lengths of prytanies were irregular. In *Calendars of Athens*, Chapter VI, we offered a text of the logistai inscription (Prytany years 426/5, 425/4, 424/3, 423/2) with regular prytanies and followed this with a critique of Meritt's fifth-century calendar in *BCH* 81 (1957) 293–300. It was in response to these two studies that Meritt claimed in Chapter 10 of *Year* that the prytanies were irregular and that Skirophorion was sacrosanct. For those who believe that the prytany calendar was regular or that tampering with the festival calendar was possible, Meritt's theory of a sacrosanct Skirophorion rests on sand.

Single dates which might be called kat' archonta are identified as early as the Choiseul Marble in the fifth century. D. Lewis, *CR*

[9]Cf. his *Athenian Year* 203.

86 (1972) 258, wrote: "It is now agreed between Pritchett and Meritt that in Mounichion of whatever year it was there was an intercalation of days followed by suppression of days in the archon's calendar, and it is something to have the possibility demonstrated for the fifth century." For a professional epigraphist the language is strange. Two scholars might agree on error. The point is that the texts are on the stone. See the table in *BCH* 101 (1977) 101, and Lewis' text in *IG* I³ (1981) 377 lines 3–10. Similarly, for the latter part of this year, B. D. Meritt, "The Choiseul Marble," *Mélanges helléniques offerts à Georges Daux* (1974) 265, wrote, "Two days were added to Thargelion between the second and the eleventh and then subtracted before Skirophorion 5." The text which I tried to decipher from the worn surface yields even more extensive tampering.

EXAIRESIMOI DAYS

The problem of exairesimoi days is the most critical one for students of the calendar and will never go away. Whereas some texts expressly give κατ' ἄρχοντα and embolimoi days, dates taken out of the calendar in compensation were not recorded. Any response to the problem of the omitted day, as well as to those of the length of months and the character of many archon years, depends primarily on what solution is suggested. Any solution is hypothetical, but some theories about limits may be examined.

The term ἐξαιρέσιμος (ἡμέρα) is defined by LSJ as (day) 'taken out of the calendar'.[10] The word is used four times in Geminos 8.55-56 in the context of days omitted in the cyclic

[10]I follow the Greek usage in referring to an omitted "day". Several pedants have objected that in English "date" should be used. The day listed on the parapegmata was not counted.

calendars of the astrologoi.[11] Heath in his translation uses the term 'eliminated'. Cicero *Verres* 2.2.52.129, uses the Greek word in his accusation against Verres for manipulating the calendar in Sicily.[12] For Plutarch's use of the verb ἐξαιρέω in connection with the omission of Boedromion 2, see Jon D. Mikalson, *The Sacred and Civil Calendar of the Athenian Year* (1975) 47. Philostratos, who composed the *Lives of the Sophists* at Athens some time in the first third of the third century of our era, wrote (558) about Herodes Atticus and his daughter Panathenais (tr. Wright): "Thus, then, his grief for Regilla was quenched, while his grief for his daughter Panathenais was mitigated by the Athenians, who buried her in the city, and decreed that the day on which she died should be taken out of the year" (ἐξαιρεῖν τοῦ ἔτους).[13]

Exairesimoi Days Before Embolimoi Ones. Meritt, "Calendar Problems," *TAPA* 95 (1964) 253–254, theorized that days might be suppressed and then rectified by subsequent corresponding additions. He wrote,

> Pritchett has now devoted a section of his "Ancient Athenian Calendars on Stone" to these "Suppressed Days," and has given testimony to his growing belief that suppression of days might indeed precede the necessary corrections to be made later by addition. It is good to have this principle recognized. It makes easier the understanding of Aristophanes' complaint about the calendar in the *Clouds* 615–16:

[11]On the basis of this passage in Geminos, Meritt advocated for a good part of his career that the 64th day was omitted in the Athenian untampered festival calendar; see, for example, his table in *Arch. Eph.* 1968 98.

[12]L. Dubois, *REG* 112 (1999) no. 641 p. 710, says Itonios, to which the term ἐπαγώγιμος is applied, was the intercalary month at Tauromenium. However, Apellaios is labeled δεύτερος in *IG* XIV.429 line 9, followed by Itonios. Apparently, epagogimos was equivalent to embolimos, a month inserted irregularly, as in Boiotia and Attika.

[13]See Simone Follet, *Athènes au IIᵉ et au IIIᵉ siècle* (1976) 358.

> ὑμας δ' οὐκ ἄγειν τὰς ἡμέρας
> οὐδὲν ὀρθῶς, ἀλλ' ἄνω τε καὶ κάτω κυδοιδοπᾶν·
>
> and it is only logical that a dispensation which would permit irregularity in one direction would permit it also in the other. This may, then, be the answer to the intercalary character of the year of Heraclitus, and there is no need to move Heraclitus away from 137/6 and in so doing disrupt the whole chronological framework of this part of the second century. The theory of an early suppression of days may now be applied to the year of Archippus (318/7).

In *Athenian Year* 206, Meritt had earlier expressed the same theory. He gives no pagination to my section in *Stone*, but the reference is to pages 342–343, where I expressed no 'growing belief', but pointed out that, whereas in *Calendars of Athens* 72, I supported such a theory to accommodate two dates given in Ptolemy *Almagest* VII.3 for observations in Egypt by Timocharis in 295/4 B.C.,[14] but I recanted when I found, after a study with van der Waerden, that the dates were not Athenian calendric dates, but Timochares and others were using the names of Athenian months in their artificial astronomical calendars. Meritt later offered different interpretations for the years in question, but his theory illustrates the numerous possibilities for recasting the progress of days in the festival calendar. Entirely apart from the debate about the character of any particular year, the logical possibility exists that the state might have the occasion to advance an event in the festival calendar, and I agree with Meritt's argument in *TAPA* 1964; cf. *BCH* 85 (1961) 21-22.[15] If the political situation

[14]For Timocharis' observations, see B. R. Goldstein and A. C. Bowen, *Centaurus* 32 (1989) 272–293.

[15]Plutarch writes (*De fraterno amore*, c. 18 = *Mor.* 489B): "The Athenians always omit the second day of Boedromion thinking on that day occurred Poseidon's quarrel with Athena." Moreover, Plutarch made the problem of why the Athenians omitted this day part of the subject of one of

required, the authorities did not hesitate to do strange things to the festival calendar (e.g., Plutarch *Dem.* 26).

Meritt devotes his last article with the title of the omitted day (*ZPE* 35 [1979] 145–151), to *IG* II^2 1072, an inscription of the Hadrianic period, in which the calendar equation, he finds, has exairesimoi days before embolimoi ones. He writes (146–147), "There is no instance in which the festival count has ever been shown to run ahead of the prytany calendar." He does not mention his own treatment to the contrary, which illustrates the fact that with Meritt it is always important to trace his many rules throughout his studies. In this case, he posits a solution of a mason's error, a mistake of ιϵ for kappa in reading his copy. S. Follet, *Athènes au II ^{e}et au IIIe siècle* (1976), treats the inscription on p. 72.

For the subtraction of days before intercalation, see n. 10 in the next chapter.

Single Festival Dates. In *Calendars of Athens* (1947) p. 17, we advanced the theory, after our catalogues of kat' archonta and embolimoi days, that single festival dates without specification might be kat' archonta ones. Only thus was it possible to assign regularity to the prytany calendar. Our predecessors had assigned to prytanies a wide variety of days in any given year. We devoted a good part of the book to tables of known equations at the time to showing the differences in the count between prytany and festival dates on the assumption that the prytanies were predetermined

---

the Symposiac questions (9,6 = *Mor.* 740F and 741B), although the text with the exception of the title is unfortunately largely lost. Although we have a date Boedromion 2 in one inscription of the fifth century B.C. (*IG* I^3 377), we have found no reference to the day in any document of subsequent centuries. Hence, it is reasonable to suppose that Plutarch may be right for later centuries. On this date, then, the festival calendar, if untampered with heretofore, would become one day ahead of the lunar calendar; cf. J. G. Frazer, *Apollodorus* (Loeb) 2 p. 79.

and fixed. Our critics through Woodhead *Agora* XVI (1997) have offered no comparable tables. Rather, they limit themselves to treating separate years, restoring equations giving embolimoi/exairesimoi days and prytanies of differing lengths. In Table 4, I reproduce from *Phoenix* 30 (1976) 350, the number of embolimoi days Meritt hypothesized in treating equations after 1968.

TABLE 4

TABLE OF SOME OF MERITT'S TAMPERED DATES

NO. OF DAYS	MONTH	SOURCE
1	Anthesterion (VIII)	*GRBS* 17 151
"one or two"	Boedromion (III)	*Hesperia* 38 (1969) 437
2 (*quater*)	Pyanopsion (IV)	*AJP* 93 (1972) 165
	Gamelion (VII)	*Arch. Eph.* 1968, 113
	Anthesterion (VIII)	*Arch. Eph.* 1968, 114
	Thargelion (XI)	*Mélanges ... Daux* 265
3 (*bis*)	Pyanopsion (IV)	*Agora: Inscriptions* 15.114
	Elaphebolion (IX)	*Klio* 52 281
4	Elaphebolion (IX)	*Hesperia* 38, 440
"four or five"	Maimakterion (V)	*TAPA* 95 (1964) 254
5	Anthesterion (VIII)	*Arch. Eph.* 1968, 114
"five or six"	Hekatombaion (I)	*Agora: Inscriptions* 15.114
6 (*bis*)	Elaphebolion (IX)	*Hesperia* 38, 108
	Elaphebolion (IX)	*AJP* 93, 168
7 (*bis*)	Elaphebolion (IX)	*Hesperia* 38, 108
8 (*bis*)	Hekatombaion (I)	*Agora: Inscriptions* 15.157
	Pyanopsion (IV)	*Agora: Inscriptions* 15.164
10	Elaphebolion (IX)	*GRBS* 17 150
11	Elaphebolion (IX)	*Arch. Eph.* 1968, 114

NO. OF DAYS	MONTH	SOURCE
13	Pyanopsion (IV)	*Hesperia* 38, 440; *AJP* 93, 165
16	Posideon (VI)	*Hesperia* 38, 440
19 or 20	Mounichion (X)	*Hesperia* 38, 440
20	Elaphebolion (IX)	*Agora: Inscriptions* 15.192
1 month + 1 day	Hekatombaion (I)	*Agora: Inscriptions* 15.113

This table illustrates the extent of tampering posited by the scholar who has provided our latest archon-table and serves as a guide to those who may wish to restudy correspondences between the festival and prytany calendars as well as the character of many years.[16] Conceding, as one must, that what Meritt has his archons do in tampering with the festival calendar, others might duplicate, one may reconstruct calendars for many years different from those of Meritt with respect to the character of years, the length of months, and the length of prytanies. Such phrases as "this adjustment is a simple one", "it is preferable to accept the reconstruction which results in the least tampering", "this reconstruction is more economical", are meaningless.[17] I offer as an example a series of Meritt's "guaranteed" texts based on a theory which Meritt himself later abandoned without a backward glance.

In *Athenian Year* p. 51, Meritt wrote that there was "a hard core which can be satisfied with nothing but forward count." He listed seven examples (pp. 52–53). Later, he says he was persuaded that the Athenians used only backwards count, but he never turned back the clock. In this context of irregularities in

[16]Never has the discovery of a new inscription with a calendar equation confirmed the arrangement of months and days previously offered.

[17]Woodhead's *Agora* XVI is replete with such phrases, which, however, he ignores when Meritt's reconstructions show greater tampering. Nowhere does he offer a table of triple dates and of embolimoi days which attest on the stones the extent of tampering.

the festival calendar, it is instructive to convert his texts to backward count.[18]

IG II² 504 (Archon Nikokles, 302/1). His restored equation, which he pronounced as 'guaranteed', is Skirophorion (XII) 29 = Prytany XII 22. By any stretch of the imagination, this equation requires tampering in the month of Skirophorion.[19]

IG II² 650 (Archon Polyeuktos, 249/8). Elaphebolion (IX) ἐνάτει μετ' εἰκάδας = Prytany IX 30. The festival calendar lagged eight days behind the prytany one. This discrepancy occurs after the Dionysia and contradicts Meritt's rule of immediate correction.

IG II² 848 (Archon Archelaos, 222/1) = Dow, *Prytaneis* 36B. Boedromion (III) 24 = Prytany IV 3. Four days were intercalated into the festival calendar. Cf. *Calendars of Athens* p. 91.

IG II² 891 and 892 (Archon Symmachos, 188/7). Meritt claimed that his restorations required forward count in two inscriptions: *Athenian Year* 154–158. The interesting feature is that Woodhead (*Athenian Agora* XVI p. 373), as recently as 1997, has tacitly endorsed Meritt's text with the two forward counts, writing, "He (Meritt) reviewed all the evidence in *Athenian Year*, pp. 154–158 (see *SEG* XXI, 430–434), and was able to devise a satisfactory arrangement of months and prytanies ...". Meritt corrected the readings of Koehler, Wilhelm, and Kircher, and may well be right, but they have not been checked. In *IG* II² 891, he offers the festival date as ἐ]νάτει <ἐπὶ δέκα>, a stonecutter's omission, read earlier as ἐνδε]κάτει. The five inscriptions of this year, all non-stoichedon, were published with the omission of restorations in *Calendars of Athens*, p. 29. One sees that the texts are so frag-

[18]Cf. *CSCA* 9 (1977) 186.

[19]There are other restorations. Suffice it to say here, the guarantee became unglued.

mentary that restorations can be only exempli gratia, including the nature of the year.[20]

Prytaneis No. 64 (Archon Philon, 178/7). Two equations: Pyanopsion (IV) ἐνάτει μετ' εἰκάδας = Prytany IV 30 and Maimakterion (V) ἕκτει = Prytany V 10. Meritt guaranteed that the first equation was with forward count and then assumed, following Dinsmoor, that ἕκτει was inscribed by error for ἐνάτει. As indicated in *Calendars of Athens* p. 85, the first equation (IV 22 = IV 30) shows a retardation of eight days; four had been corrected by the time of the second (V 6 = V 10). In a period of ten days, four had been taken out of the festival calendar, if ordinary. Four days were still to be exairesimoi. The example is one of gradual compensation.

IG II² 953 (Archon Tychandros, 160/59). There are two equations for this year, which were studied in *Calendars of Athens* pp. 28 and 76. We substituted ε[ἰκοστεῖ for ἐ[νάτει in *IG* II² 953, giving an equation for an intercalary year with prytanies of 32 days: Posideon 29 (δευτέραι μετ' εἰκάδας) = Prytany VI 20 = 180th day. The second equation in *Inscriptions de Délos* IV no. 1497 *bis* gives [Elaphebolion] (IX) 4 = Prytany IX [1]9 = 275th day.[21] The first equation involved a retardation of four days, the

[20]In the five texts, we have two prytany numbers, three festival months, three festival dates, and two prytany dates. No single inscription has more than two elements. One would have to check the stones for spacing. For my treatment of the calendar according to pre-Meritt readings, see *Calendars of Athens* 84–85.

[21]The thrust of Meritt's argument on p. 163, which is frequently repeated, is that it is absurd to claim that the Athenians had a lunar calendar and at the same time to allow tampering to extend past a new moon. Actually, in this case we left the matter open whether the retardation was continuous or whether the calendar was corrected and then retarded for a second time, although I now favor the former. Meritt's objection was to violating any "date of new moon for any month anywhere in the year", a complete misunderstanding of the practice of tampering. In *Agora* XV no.

second of five. Meritt did not treat the equations until his *Athenian Year* 162–163, when he conceded the second retardation of five days, but argued vehemently in favor of the first equation as Posideon δευτέραι μετ' εἰκάδας = Prytany VI 9. If this is taken as backward count, we have Posideon 29 = Prytany VI 9, which means that there had been seven exairesimoi days in the festival calendar in the first half of the year. It follows that by the time of the second equation there must be twelve embolimoi days.

I shall return to the subject of tampering in treating the prytany calendar. Our concern here has been only with irregularities in the festival calendar. Suffice it to say that what one archon might do, another might do similarly.[22]

REASONS FOR EMBOLIMOI/EXAIRESIMOI DAYS

Ancient authors provide evidence abut the shifting of calendar dates to meet emergencies and the intercalation of days by naming each successive one the same. Thucydides, in 5.54, tells the story that the Argives set out on τετράδι φθίνοντος (27) of the month preceding the Karneian festival to invade Epidauros and continued to observe that day during the whole time they ravaged it (ἄγοντες τὴν ἡμέραν ταύτην πάντα τὸν χρόνον). Allies of Epidauros could not help because the month of the Karneia was sacred. Dover (*HCT* IV.75) claims that the device also served to satisfy the piety of the Argives, but this is not in Thucydides.

---

121 p. 114, he posits a span of retardation of three days to extend over "about six weeks."

[22]Some texts are based on restorations. Thus, in *Hesperia* 38 (1969) 440, Meritt's "elapsed days" count calls for sixteen days for an ordinary year (223/2), whereas new restoration of this non-stoichedon inscription in *Agora* XV no. 128 p. 120 allows for no elapsed days in an intercalary year. In a text which he dates in 298/7 (*Hesperia* 38 [1969] 107), he assumes retardation of seven days and obtains an equation for an intercalary year instead of an ordinary year as previously interpreted.

Plutarch (*Alex.* 16)[23] cites a similar expediency on the part of Alexander in doubling the month.

According to Xenophon *Hell.* 4.7.2–3, 5.1.29, the Argives juggled the calendar in the Korinthian War. They proclaimed a sacred truce every time the Spartans invaded the Argolid by dispatching heralds with wreaths and the proper insignia to warn them off their territory. So compelling was the religious sentiment that the Spartan King Agesipolis went both to Olympia and Delphi to get the approval of the oracles not to submit to such a pretense.

The tour de force by which Stratokles humored the whim of Demetrios in which Mounichion (X) was declared to be Anthesterion (VIII) for initiation into the Lesser Mysteries and then decreed to be Boedromion (III) for the august Eleusinian Mysteries is told by Plutarch *Dem.* 12; cf. Diodoros 20.110.1; A. G. Woodhead, *Hesperia* 58 (1989) 297–301. Similar manipulations to accommodate foreign potentates may have become not uncommon in Hellenistic times.[24]

The prime epigraphical example of embolimoi days is found in a law of the Euboian Confederacy about 290 B.C. (*IG* XII.9.207; *IG* XII Suppl. p. 178; *SEG* 30.1055; *SEG* 34.896) which prescribed that Dionysiac troupes were to appear at specified dates in four Euboian cities. In connection with the appearance of the troupes on the dates designated for the local Dionysia, the prescription was given that the calendars might be adjusted for up to three days (lines 28–29). Chapter VII of the second edition of A. Pickard-Cambridge, *The Dramatic Festivals of Athens* (1968) by Gould and Lewis is devoted to the organizations of

[23]For powerful religious convictions which outweighed military and political considerations, see *War* 3.69–70.

[24]Cicero (*Verres* 2.2.52.129) charged that Verres declared the Ides of January to be the kalends of March. Tricks played with the calendar at Rome for political authorities for party purposes have not been collected.

dramatic and musical performers at Athens and elsewhere. Athens, as the first home of plays, was visited by famous actors, whose travel plans had to be accommodated, just as the Dionysiac guild on Euboia.[25] The tragic actor Athenodoros was fined by the Athenians for failing to appear at the Dionysia, and the fine was paid by Alexander: Gould and Lewis p. 280 n. 7. *IG* II^2 1330 (= Dittenberger, *OGIS* 352; *SEG* 45.121, 2324) is a lengthy decree of the guild of the technitai of Dionysos who celebrated a festival for the birthdays of the Cappadocian king Ariarathes V and his wife Nysa in Athens on Metageitnion 14 and 15 (lines 33–34). This was not a state festival, but the catalogue of musical and dramatic events suggests that it was better attended than many other festivals. Such celebrations would rank with state festivals as possible candidates for tampering. One of the embolimoi dates indicates that the calendar had been stopped on the 9th of Elaphebolion (271 B.C.) for five days: Ἐλαφηβολιῶνος ἐνάτει ἱσταμένου τετάρτει ἐμβολίμωι. Gould and Lewis write (p. 65): "This procedure is unintelligible except as a means for gaining time to complete the preparations for the Dionysia." They suggest that the 10th was the day of the πομπή. For this same intercalation of four (or more?) days, Dinsmoor, *Hesperia* 23 (1954) 309, wrote:

> Evidently the preparations for the Dionysiac festival of 271/0 B.C., presumably celebrated with special splendor as shown by the fact that the *aganothetes* Thrasykles rebuilt the upper part of the choragic monument of his father Thrasyllos to commemorate it, consumed more time than had been allowed, and the four extra days had to be inserted in order that the *pompe* might still be held on Elaphebolion 10.

[25] A truce for the Mysteries lasted fifty-five days from Metageitnion 15 to Pyanopsion 10: *IG* I^3 6.B. Various states including the Islands and Egypt sent worshippers: G. Mylonas, *Eleusis and the Eleusinian Mysteries* 244. On occasions, delays may have been accommodated, if caused by rough seas and the like.

According to Meiggs and Lewis, *GHI* Revised Ed. p. 221, the intercalation of a second Hekatombaion was "presumably in order to give longer notice of the date at which first-fruits must be delivered at Eleusis." In *Hesperia* 52 (1963) 437, Meritt suggests that the calendar was retarded at a time having nothing to do with a festival "for diplomatic negotiation with Aitolia for defense against anticipated war."

We have no clue to the specific dates of the majority of Athenian festivals. Thus, the Ptolemaia, which Parker, *Athenian Festivals* 274–5, calls the "most elaborate and long-enduring festival introduced in the third century", is not dated in Deubner. An up-to-date paper organizing the evidence for the dates of Athenian festivals would be a useful contribution. For festivals, Athens was an open-air society, as illustrated by the theater of Dionysos. Processions and various contests were exposed to the elements. In a valuable article, J. D. Mikalson, *GRBS* 23 (1982) 213–221, catalogues various religious activities attested in the literature for Athens. Of processions, he writes (p. 217):

> Processions are attested for the Bendideia (Pl. *Resp.* 327A), Brauronia (Philochorus *FGrHist* 328F101), City Dionysia, Choes, Delia (*I. Délos* IV 1869), Eleusinia (*IG* II^2 930.8), Eleusinian Mysteries, Haloa (*Anecd. Bekker* I 384.31), Panathenaia (Parthenon frieze), Pyanopsia (Harp. *s.v.* πυανόψια), Skira (Harp. *s.v.* Σκίρον), Thesmophoria (Isae. 6.50), and Thargelia (Arist. *Ath.Pol.* 56.26).

Earthquakes,[26] eclipses, periods of bad weather,[27] meteorological omens might cause postponements. There are miscellaneous clues which show that a superstitious public viewed meteorological signs with alarm and might stop the business at hand, but nothing

[26]In Athens about ten years ago, an earthquake caused thousands to flee the city and interrupted business for several days.

[27]On uncertain rainfall in Athens, see Robin Osborne, *Classical Landscapes* (1987) 31–34.

specific about festivals. The disastrous effect of the eclipse recorded in Thucydides 7.50.4 is well attested. The Ravennas scholiast to Aristophanes *Acharnenses* 171 says, "The Athenians were careful to mark any signs derived from the sky and when such a sign occurred they would break up the meetings of th ekklesia if they were on the point of executing [lacuna] or anything else." Pollux (8.124) explains that the signs were interpreted by exegetai: ἐξηγηταὶ δ' ἐκαλοῦντο οἱ τὰ περὶ τῶν διοσημειῶν καὶ τὰ τῶν ἄλλων ἱερῶν διδάσκοντες. When a sign occurred, the public assembly at once broke up and the law courts ceased to sit (Pollux: ἀνίστατο δὲ τὰ δικαστήρια, εἰ γένοιτο διοσημεῖα). Thus, in 420 B.C., the Athenians were on the point of making an alliance with the Argives and their confederates. "But," says Thucydides (5.45), "before the final vote was taken an earthquake happened, and the assembly was adjourned." Aristophanes in the *Ecclesiazusae* (791–793) mentions as reasons for not carrying a decree (tr. of Rogers): "Why, if perchance an earthquake came, or lightning fell, or a cat crossed the street, they'll soon cease bringing in, you blockhead you!" The chorus of the *Nubes* (577ff.) takes credit for saving the Athenians: "For though we benefit the state most of all the gods, to us alone you do not offer sacrifices or pour libations, who watch over you. For if some rash expedition is afoot, then we thunder or drizzle. Again, when you were bent on choosing as your strategos that Paphlagonian tanner, we contracted our brows, and made a dreadful noise overhead." Postponement of public business is reported in a passage in the *Acharnenses* (171) where Dikaiopolis waxes indignant with the Thrakians: "I warn you not to hold the assembly about the Thrakians' pay. I tell you there's a portent come: I felt a drop of rain." Hereupon the herald terminates the meeting: "Thrakians begone, come back in two days' time. The assembly is dissolved."

A possible area of speculation about modification of the calendar is that of agriculture. A. C. Brumfield, *The Attic Festivals of Demeter and their Relation to the Agricultural Year* (New York 1981) claims that the Thesmophoria in Pyanopsion was associated with the sowing of winter cereals (p. 88); cf. *OCD*³ 1509. She also suggests (p. 108) that the Haloa in Posideon was a festival for the pruning of the vine and the tasting of stored wine. Any festival tied to the soil is subject to modification because of the weather.

One may think of inclement weather, but this will not usually explain retardations of more than a day or two in the Athenian climate, and that at the time of the festival. Meritt has offered reconstructions of the calendar of several years by hypothesizing that retardations were common in Boedromion (Eleusinia) and Elaphebolion (Dionysia),[28] but many of his reconstructions should be labeled exempli gratia and are no better than those he would replace. Deubner says the Dionysia fell on Elaphebolion (IX) 9–13. Of the three attested cases of kat' archonta dates in this month (Nos. 9, 10, and 19 in Table 1), all are later than the festival and the retardation in the case of No. 10 is 20 days (±). If the retardation had anything to do with the Dionysia, it was long uncorrected. Of his theoretical calendars in Elaphebolion, Meritt writes (*Year* p. 148), "I assume that the compensation by omission of days was made after the festival but before the end of the month." Page 208, "Our own study has shown that most of the additions in which three or more days were involved came at the time of the City Dionysia." He does not share his "own study" with the reader. Any future student of a new archon-table has a great task in separating the chaff from the seed. This is particularly true of the calendric sections of Woodhead's *Agora* XVI.

[28]See Meritt/Traill, *Athenian Agora* XV p. 23.

USE OF FESTIVAL CALENDAR.

Geminos, in his *Eisagoge* (8.6–9), explained the purpose of the Greek calendar as keeping sacrifices in the same months from year to year (tr. Thomas Heath, *Aristarchus of Samos* [Oxford 1913] 284-285):

> 'The ancients had before them the problem of reckoning the months by the moon, but the years by the sun. For the legal and oracular prescription that sacrifices should be offered after the manner of their forefathers was interpreted by all Greeks as meaning that they should keep the years in agreement with the sun and the days and months with the moon. Now reckoning the years according to the sun means performing the same sacrifices to the gods at the same seasons in the year, that is to say, performing the spring sacrifice always in the spring, the summer sacrifice in the summer, and similarly offering the same sacrifices from year to year at the other definite periods of the year when they fell due. For they apprehended that this was welcome and pleasing to the gods. The object in view, then, could not be secured in any other way than by contriving that the solstices and the equinoxes should occur in the same months from year to year. Reckoning the days according to the moon means contriving that the names of the days of the month shall follow the phases of the moon.'

We have two passages which claim to give Sokrates' view of the education of the young in the arranging of the calendar and the appointment of days for monthly and annual sacrifices and feasts. Plato in *Laws* 7.809c writes (Loeb tr.):

> ... also arithmetic, of which I said that there ought to be as much as everyone needs to learn for purposes of war, house-management and civic administration; together with what it is useful for these same purposes to learn about the courses of the heavenly bodies—stars and sun and moon—in so far as every State is obligated to

> take them into account. What I allude to is this—the arranging of days into monthly periods, and of months into a year, in each instance, so that the seasons, with their respective sacrifices and feasts, may each be assigned its due position by being held as nature dictates, and that thus they may create fresh liveliness and alertness in the State, and may pay their due honours to the gods, and may render the citizens more intelligent about these matters.

Xenophon *Mem.* 4.7.4-5 states (Loeb):

> Similarly he recommended them to make themselves familiar with astronomy, but only so far as to be able to find the time of night, month and year, in order to use reliable evidence when planning a journey by land or sea, or settling the watch, and in all other affairs that are done in the night or month or year, by distinguishing the times and seasons aforesaid. This knowledge, again, was easily to be had from night hunters and pilots and others who made it their business to know such things. But he strongly deprecated studying astronomy so far as to include the knowledge of bodies revolving in different courses, and of planets and comets, and wearing oneself out with the calculation of their distance from the earth, their periods of revolution and the causes of these. Of such researches, again he said that he could not see what useful purpose they served. He had indeed attended lectures on these subjects too; but these again, he said, were enough to occupy a lifetime to the complete exclusion of many useful studies.

One would like to know more about the popular lectures.

Since the Athenians had both a festival and a prytany calendar, the question arises about their comparative use. J. A. Notopoulos, "The Conciliar and Civil Calendar in I.G., I^2, 324," *AJP* 66 (1945) 411–414, addressed the problem, observing that all loans were made in terms of dates in the prytany calendar, but two festival dates were added in those made by the Other Gods, suggesting that the Other Gods had conservative, rural cults, and the two dates were added to accommodate their records. The fact that an inscription (*SEG* 34.95) recording a decree of the ekklesia with the inventory of a sanctuary of Aphrodite (?) gives kat' archonta and kata theon (i.e. festival) dates might support this suggestion.

I note a few, sometimes conflicting, clues about what administrative bodies used which calendar. The Areopagos tried homicide cases on the two or three ἀσέληνοι days, or nights, at the end of the lunar month,[29] a practice which must have arisen long before there was a prytany calendar. This suggests that the courts continued to use a calendar which was lunar.[30] In Demosthenes 21 *Meidias* 86 with scholion, the last day on which the arbitrators met was determined by the festival calendar.[31] The temple of Dionysos in Limnai could be opened only once a year, namely on 12 Anthesterion (Demosthenes 59 *Neaira* 76) and so on. According to Demosthenes 42 *Phainippos* 5, legal proceedings occurred on Metageitnion (II) 5. The last day of the month was the one fixed for the lodging of deposits by intending litigants; hence Strepsiades' dread of it in Aristophanes *Clouds* 1134, 1180, 1196–7. See A. H. Sommerstein's notes on p. 216 of his edition. Plutarch (*Mor.* 828A) has it that the noumenia was made accursed and detested by the money-lenders, when interest was due or overdue; cf. Horace *Satires* 1.3.87.

Margaret Crosby, in *Hesperia* 10 (1941) 14–30, published a complete stele which is the poletai record of the year 367/6 B.C. Lines 8–39 give the account of the sale of a house and of the

[29]For these two, or three, moonless nights, see *War* 3 p. 211. See also *ZPE* 128 (1999) 88. Jon D. Mikalson, *The Sacred and Civil Calendar* (1975) 22–23, has shown that the Areopagos did not try cases every month (for example, in Hekatombaion).

[30]The informer in Plutarch *Alk.* 20.5 was convicted of perjury, who stated that he recognized by the light of the moon the faces of the Hermokopidai, whose offense was committed ἕνης τε καὶ νέας οὔσης.

[31]Douglas MacDowell, in his 1990 commentary p. 310, says that the two lunar months should not be attributed to Demosthenes, but the reference to the last day is to the lunar calendar.

three claims that were brought against it during the course of the sale. The date of the registration of the house is given as Mounichion 10. Lines 40–83, however, are devoted to the leasing of mining concessions. Here the dates are given prytany by prytany (1, 2, 3, 4, 5, 7, 9). A public debtor who had not paid by the ninth prytany lost his civic rights (Demosthenes 24 *Timokrates* 87). The epistatai οἱ Ἐλευσινόθεν, *IG* II[2] 1672, itemized payments according to prytanies in the year 329/8. In *IG* II[2] 1673 (333/2 ?), for which Clinton has given a new text in *AE* 1971. 83–113 (= *SEG* 34. 122), salaries were paid by prytany year, but some items are listed by festival dates: Glotz, *REG* 31 (1918) 210. In the period of ten phylai in 345 B.C., the two obols per diem pay (Aristotle *Ath.Pol.* 49.4) for the disabled was made according to the prytany calendar (τὸν τῆς πρυτανείας μισθόν), not the festival one: Aischines 1 *Timarchos* 103-104. The scholiast on this passage writes: καὶ μετροῦσι κοινότερον οὐ πρὸς μῆνας ἀλλὰ πρὸς πρυτανείας τούς τε μισθοὺς καὶ τόκους καὶ ἐνοίκια. This is tantamount to saying that it was the financial year. The stability of the prytany calendar may be explained in part by its use in financial records. Aristotle, *Ath.Pol.* 47.2–4, shows that dating by prytanies was used by πωληταί for their financial dealings, which included the leasing of mines (M. Crosby, *Hesperia* 19 [1950] 192), and by the βασιλεύς in the letting out of sacred τεμένη; see George Huxley, *AJP* 86 (1965) 302.

It seems only natural in a state which set aside so many days to some form of religious devotion that a calendar regulating these acts was instinctively followed by the ordinary citizen. This is illustrated by the fact that both Aischines and Demosthenes refer to meetings of the ekklesia in terms of the festival calendar. Similarly, in dealings with citizens of foreign states which followed a lunar calendar only, the Athenians might use festival dates. See, in particular, *IG* II[2] 951 add., as discussed in *ZPE* 128 (1999) 83–84. In Lysias 23.6, we are told that the fresh-cheese market was held on the last day of the festival month. A marriage was concluded when Polyzelos was archon (366) in Skirophorion, and the divorce document written when Timokrates was archon (364) in Posideon (Demosthenes 30 *Onetor* 1 15), a clue that suggests that the festival calendar was used for general indications of time.

As to the kata theon calendar, I suggest that the occurrence of the phrase in some second-century legal documents is more meaningful than merely to indicate the degree of tampering with the count of days. The populace must have been widely aware of the two festival calendars, kat' archonta and kata theon. The kat' archonta calendar certainly reflected the major festivals attended by the general populace in Athens, but the demesmen (Marathon, Erchia, etc.) with their calendars calling for the sacrifice of one pig or one goat on any given day, attended by a local audience, may have abided by the kata theon days except on the state occasions.[32] When Meritt (*Hesperia* 32 [1963] 437) finds that one month had 42 days and another 17, I question whether such tampering was reflected in observing the sacrificial calendars of country people. This is only speculation, but one must concede that the people of Attika were well aware of the two festival calendars. We need enlightenment about how heralds informed country people in remote demes about the archon's intentions with regard to embolimoi/exairesimoi days. Charles Dickens, in *Barnaby Rudge*, a historical novel set in the 1780's, presents a picture of the streets of London, and in one sentence writes: "Then there was the watch with staff and lantern crying the hour, and the kind of weather; and those who woke up at his voice and turned them round in bed, were glad to hear it rained, or snowed, or blew, or

[32]For the five sacrificial calendars, see *ZPE* 128 (1999) 93. For a good discussion, see J. D. Mikalson, "Religion in the Attic Demes," *AJP* 98 (1977) 424–435. He shows that demesmen traveled to Athens and attended major state festivals such as the Panathenaia and Dionysia, but each deme had its own calendar for many rather simple sacrifices.

froze, for very comfort's sake." The role of heralds who conveyed the progression of days to the widespread populace of Attika, embracing a territory of 930 square miles (= 2,400 sq. kil.), must have been demanding. For a catalogue of the Heralds of the Boule and Demos, see Meritt/Traill, *Agora* XV pp. 14–15, but they do not discuss their function. For κήρυκες, see Oehler, *RE* s.v. Keryx (1921) 349-357. A. S. F. Gow, *Machon* (1965) 86, discusses a fragment of this comic poet which parodies the conduct of the wit Stratonikos at Abdera about the proclamation of the new moon.[33] We hear of νουμηνιασταί at Athens (Lysias frg. 53), but they were apparently a society meeting on the noumenia.[34]

Frequency of Festivals. A number of isolated passages, spaced over the centuries, testify to the frequency of festivals and holidays at Athens. The festivals, apart from their religious and ritual side (processions, sacrifices, banquets, and torch-races), included gymnastic, equestrian, musical, and dramatic contests.

In *Ath.Pol.* 3.2, Pseudo-Xenophon makes the general statement that it is difficult to negotiate with the boule and ekklesia, continuing, "For how could they do this? First of all they have to hold more festivals than any other Greek city (and when these are going on it is even less possible for any of the city's affairs to be transacted) ..."; ibid. 3.8 ("twice as many ..."). E. Kalinka, *Die pseudoxenophontische Athenaion Politeia* (Leipzig 1913) 264, pointed out that the language of *Ath.* 3.2 (ἧττόν ... δυνατόν) means that on occasions public debate might take place on a state

[33]There were also societies of the Tetradistai, Dekadistai, Eidakeis, all named from the days of the month; see R. A. Parker, *Athenian Religion* 335–336. Did they meet according to the kat' archonta or kata theon dates?

[34]For the proverb about the heralding of the new moon, see Leutsch and Schneidewin, *Corpus paroemiographorum graecorum* 1 App. 2 no. 61.

holiday. In 2.38, Thucydides has Perikles say (Loeb): "Moreover, we have provided for the spirit many relaxations from toil: we have games and sacrifices regularly throughout the year and homes fitted out with good taste and elegance; and the delight we each day find in these things drives away sadness." Cf. Sophokles *OC* 1006–1007 εἴ τις γῆ θεοὺς ἐπίσταται τιμαῖς σεβίζειν, ἥδε τοῦθ' ὑπερφέρει. In Plato *Alkibiades* II.148E, the point is made that the gods preferred the Spartans to all the offerings and observances of the Athenians (Loeb):

> ... for we "—such was the message—" offer up to them more and finer sacrifices than any of the Greeks, and have adorned their temples with votive emblems as no other people have done, and presented to the gods the costliest and stateliest processions year by year, and spent more money thus than all the rest of the Greeks together. But the Spartans have never taken any such pains, and indeed are so neglectful in their behaviour to the gods, that they make a practice of sacrificing defective victims, and generally are very much behind us in the honours that they pay, though the wealth they possess is quite equal to that of our city.

A scholiast to Aristophanes *Wasps* 663 reckons that a sixth of the year was given over to religious festivals (εἰς ι´ μῆνας λογίζεται τὸν ἐνιαυτὸν, ὡς τῶν β´ ἑορτὰς προχωρούντων).[35] Demosthenes, a fiscal conservative, repeatedly reproved his countrymen for their

[35]To some modern Americans the number may seem small. Strong labor unions and many federal agencies limit the working week to four days, others to 40 hours. In addition, many Mondays are holidays. Various nationalities take 52 Sundays or Sabbaths, and some festivals (Easter, Passover, Ramadan, Christmas, etc.) last for several days. In the Roman calendar, Mommsen (*CIL* I [1863] p. 373) calculated the year contained 45 *dies fasti*, 194 *dies comitiales*, 48 *dies nefasti* or *feriati*, 57 *dies religiosi*, 8 *dies interdici*, and 3 *dies fissi*. On *Dies nefasti*, no business could be done because the day was sacred to some festival. *Dies religiosi* were unlucky days.

devotion to spectacles, processions, and sacrifices to the neglect of affairs of state.[36] In the Essay on the Glory of Athens attributed to Plutarch (*Mor.* 348C ff.), the author dilates on the grandeur of entertainments and relates an opinion expressed by a Lakedaimonian that the Athenians lavished on spectacles sufficient for the equipment of large fleets. A number of passages of similar import is collected by A. Boeckh, *Die Staatshaushaltung des Athener*[3] I (1886) 265 ff. Some time after 164 B.C., in a work attributed to Dikaiarchos, a pupil of Aristotle, the author wrote (C. Müller *GGM* I.97–99, paraphrase of Frazer, *Pausanias* I.xliii):

> Amusements and recreations; many holidays and a constant succession of spectacles;—all these the visitor will find in Athens. The products of the country are priceless in quality but not too plentiful. However, the frequency of the spectacles and holidays makes up for the scarcity to the poorer sort, who forget the pangs of hunger in gazing at the shows and pageants. Every artist is sure of being welcomed with applause and of making a name; hence the city is crowded with statues.

Athenaios devotes a long passage from Theopompos as to why distinguished men preferred to live outside of the city of Athens, concluding (12.532d, tr. Gulick), "The populace as a whole squandered more money on the public banquets and distributions of meat than on the administration of the State." Philostratos (*Vit. Apoll.* 4.17) says at the time of the Mysteries there was no city in Greece more populous than Athens. R. Parker, *Athenian Religion* (1996), treats festivals of Hellenistic times on pages 267–275, citing the third-century traveler Herakleides. He refers to the Ptolemaia, Panathenaia, Dionysia, and the Eleusinia as "the four

[36]Cf. *Third Ol.* 31–32; *First Phil.* 35: "How do you account for the fact that the Panathenaic festival and the Dionysia are always held at the right dates ... at the festivals everything is ordered by statute." Cf. 18 *Crown* 86, 216. Isokrates 4 *Panegy.* 43–46, dilates on the benefits of festivals.

showcase festivals". See also Volkmann, *RE* s.v. Ptolemaia (1972) 1586.[37] Endowments for festivals in Roman times are treated by J. Larsen, T. Frank's *Economic Survey* IV (1938) 362–363, 490–491. There were several at Delos, but also at Athens. Demosthenes (20 *Leptines* 21) asserts that there were perhaps slightly more than sixty recurrent liturgies (ἐγκυκλίους λητουργίας) every year in Athens. Liturgies for festivals are studied by J. R. Davies, *JHS* 87 (1967) 33–40. John Day, *An Economic History of Athens under Roman Domination* (New York 1942) 37, comments that a striking phenomenon of the second century was the holding of a large number of international festivals throughout the Hellenistic world and that festivals constituted a great expenditure of the Athenian state. Visitors and competitors came from widely separated places. P. Graindor, *Athènes sous Hadrien* (Cairo 1934) 181–182, even says that life at Athens was one continuous festival.

CONCLUSIONS

Embolimoi days were inserted into the festival calendar throughout the year and the retardation might continue over a lengthy period.

Posideon (VI) was not always the intercalary month.

No civic activity could take place on exairesimoi dates, which apparently might be subtracted before intercalations.

The reason for irregularities is conjectural, but some are judged to be related to the major festivals which included processions and

[37] In the fifth-century pseudo-Xenophon *Ath.Pol.* 3.4, reference is made to the Dionysia, Thargelia, Panathenaia, Promethia, and Hephaistia.

elaborate preparations. Few of even the showcase festivals are dated.[38]

A query has been raised whether minor sacrifices and the like in remote demes may have followed the kata theon calendar rather than the kat' archonta calendar of major festivals in the city.

[38]In assessing divergencies between the festival calendars of Athens and other city-states, one must bear in mind that all had festivals which would be subject to the same conditions for tampering.

II

OMITTED DAY: EPIGRAPHICAL EVIDENCE

The problem of the omitted day, I tied to that of the epoch of the day, a knotty problem which was treated in *ZPE* 49 (1982) 260–266 and 128 (1999) 79–93. The subject, rather neglected by early chronologists, has recently received an immense bibliography and is connected and indeed antecedent to that of the question of whether the Athenians had a true lunar calendar. This chapter treats the epigraphical evidence, the next chapter the literary.

In reviewing the subject, I have looked in vain in four prominent works on the Athenian Calendar: B. D. Meritt, *The Athenian Calendar in the Fifth Century* (1928), W. B. Dinsmoor, *Archons of Athens* (1931), Meritt, *Athenian Financial Documents* (1932), Dinsmoor, *The Athenian Archon List* (1939), for any mention of the omitted day in a hollow month. At least, the subject is not treated in any of their references to day, month, and the like in their indices, nor elsewhere that I can find. For earlier chronologists, there was little firm discussion because they found confusion (1) over the use of both φθίνοντος and μετ' εἰκάδας in the count of days in the last decade of the month, and (2) whether the count was forwards or backwards, or both. The bibliography is in August Mommsen, *Chronologie* (1883) 120–123. He, for example, opted (p. 123) for a forward count with the omission of ἐνάτη μετ' εἰκάδας (Day 29). Unger, *Philologus* 38 (1879) 442, citing Boeckh (non vidi), would omit δευτέρα φθίνοντος (also Day 29). Agreement on the 29th day was because of the literary evidence.

In *Calendars of Athens* (1947), the theory was advanced that the Athenians had a lunar calendar with years of 354±1 and 384±1 days, similar to other states, with the omitted day in the hollow month being the 29th, whether designated as δευτέρα φθίνοντος, δευτέρα μετ' εἰκάδας, or written κθ'. Ambiguity existed only for the last two days of the month. Our theory was presented on pp. 23 ff. Meritt responded by criticizing almost all elements of our theory, but focused in almost every publication subsequently on the problem of the omitted date in order to support his position of alternating full and hollow months with leap days, a theory originally advanced without any documentation. In the course of the debate, Meritt, often contradicting himself with periods of overlap, has advanced four different dates for the omitted one: 21st, 22nd, 29th, and every 64th day. All theories were presented in the context of supporting calendar equations in Athenian inscriptions, usually fragmentary, which illustrates how illusory is the interpretation of this type of evidence.

MERITT'S FOUR THEORIES OF OMITTED DAY

Two of Meritt's theories about the omitted date were later abandoned by him, but are presented briefly. The four dates illustrate serious contradictions, including interpretations of the literary evidence. One result of this survey will demonstrate that methods used to support a theory about the omitted day which were later proved erroneous are the same methods used to continue to support the same theory.

1. Omission of every 64th day

In several publications over a period of about a decade, Meritt argued that the kata theon dates were Metonic ones, in contrast with the kat' archonta dates which he claimed omitted the 21/22 day. His 'Metonic' dates are those in the 6940-day nineteen-year cycle attributed to οἱ περὶ Εὐκτήμονα καὶ Φίλιππον καὶ Κάλλιπον ἀστρολόγοι in Geminos 8.50–56. In this cyclic calendar, all months were of 30 days with every 64th day being omitted (8.56). In *Arch. Eph.* 1968. 98, Meritt devotes the better part of a page to charting "The Cycle with every 64th day indicated for omission." Preceding the table, he writes, "Year in and year out the festival calendar of the Athenians must have been much the same as the true astronomical calendar of Meton's cycle, showing at most only slight variations, if any. It would be perverse to believe that it was out of step with the phases of the moon, on which both calendars in their schematic form depended, all the time, or even, for that matter, most of the time. During the second century the Athenians thought it worth while at uncertain intervals to be specific about the true month date κατὰ θεόν on which the prytanies depended."[1] Again on p. 101, he writes, "Yet a crucial question remains, on which there is no evidence contrary to Geminus's explicit statement except what can be gleaned, possibly, from the inscriptions. Was the κατὰ θεόν calendar as applied in Athens one that omitted every 64th day, no matter where it fell within the month, or was the κατὰ θεόν calendar modified sufficiently to accord with the Athenian practice of allowing each hollow month to omit that day with which backward count in the last day of the month began?" For Meritt, the answer to the question was the former. "There is a strong case from the epigraphical record that it was." "The conciliar year of the twelve prytanies was scaled to the Metonic year."[2] Etc.

First, there are three questions to raise about the Geminos 8.56 passage. 1. The credibility of the passage. Neugebauer, Toomer, and other eminent authorities have maintained that such a calendar omitting days haphazardly in the middle of months was unworkable for the astronomical purposes for which it was ostensibly invented. For a recent discussion, see A. Jones, "New

[1]Cf. Meritt, *TAPA* 95 (1964) 235, "The astronomical calendar *par excellence* was Meton's 19-year cycle, introduced in Athens in 432 B.C. This could with equal propriety be called κατὰ θεόν." P. 260: "We have been able to show with great probability that the κατὰ θεόν calendar was the Metonic cycle."

[2]This statement is not true for the kata theon dates.

Callippic Dates," *ZPE* 129 (2000) 155–156.[3] 2. Did the count begin with the first day of each cycle? 3. Was the 64-day count progressive throughout all the cycles?

Meritt ignores No. 1 completely, but answers affirmatively to No. 2. He hypothesizes that the 19-year cycle was 0 1 0 0 1 0 0 1 etc., has Posideon as the only intercalary month, and has years of 354, 355, 384 days, none of which is in Geminos. In his Elaphebolion, for example, Day 16 (in Year 3), 18 (in 7), 12 (in 8), 30 (in 10), 14 (in 12), 8 (in 13), 10 (in 17), 4 (in 18), and 28 (in 19) were omitted. Some omitted days fall on days attested for festivals. Since the years in Meritt's so-called Metonic cycles throughout his research do not correspond to a fixed sequence,[4] in one case in the fifth century having four ordinary years in succession (also in *Athenian Year* [1961] 236), it follows that for Meritt's theory two archons might be in office in the same festival month, one for his kat' archonta festival calendar, the other for the kata theon calendar, which was the untampered festival calendar; and so with the boule. As noted in Chapter I (p. 15), an archon went into office in a month belonging to another archon. See Meritt, *Arch. Eph.* 1968. 99, where his month Skirophorion of Nikosthenes' year runs concurrently with Hekatombaion of Pelops' year, giving the approximate equations: Nikosthenes' Skirophorion 1 = Pelops' Hekatombaion 1 and Pelops' Hekatombaion 1 kat' archonta = Pelops' Metageitnion 1 kata theon. Thus, religious officials were celebrating festivals in Pelops' Hekatombaion, while civic officials including the prytaneis and councilors lacked Hekatombaion dates and had to

[3]G. Aujac, in her note to 8.56 in the Budé Geminos, writes, "Cette procédure devait compliquer singulièrement la tâche des faiseurs de calendriers: rien d'étonnant qu'un tel système ne soit jamais passé dans la pratique."

[4]See, for example, *Historia* 26 (1977) 170 ff.

begin their year on Metageitnion 1.[5] We can be sure that no archon ever celebrated the festivals of Skirophorion in the first month of the year, and no archon ever omitted the festivals of Skirophorion at the end of the year. Similarly with the prytany calendar, they ran in consecutive numerical order. The ekklesia in Prytany 1 avoided meetings on the festival dates of the Panathenaia, not those of Skirophorion. One may juggle months on paper, but the festival calendar proceeded in order throughout the seasons. Later, in presenting his final archon-table, Meritt inserted an archon, Achaios (166/5), between Nikosthenes and Pelops, but exasperated the confusion. He wrote (*Historia* 26 [1977] 165), "The final month Skirophorion of the archonship of Nikosthenes became the first month of the year of Achaios ... The year 166/5 had therefore only twelve months (Skirophorion 1 to Thargelion *ultimo*). We come to the year of Pelops (165/4) which must have had in fact thirteen months: the last month properly belonging to Achaios and his own twelve months of an ordinary year." As Meritt here separates a festival calendar (kat' archonta) from the calendar which regulates it (kata theon), placing their months in different years, so he maintained for many years that his alleged "Metonic" calendar of 30-day months with omission of Day 64 regulated a calendar omitting Day 22. In the 1977 article (p. 162), Meritt gave up his lip-service to the strange, impractical calendar in Geminos 8.56, but without a backwards glance.

After Meritt had completed his studies, S. Tracy devoted an article to the date of Achaios (*AJAH* 9 [1984] 43–47) and in his *Attic Letter-Cutters* (Berkeley 1990) 57, would date him "ca. a. 190 a." Traill, *ZPE* 103 (1994) 112, discusses Tracy's stone-

[5]It was a fundamental tenet of Meritt (and Woodhead) for his study of the prytany calendar that Skirophorion was sacrosanct in the festival calendar—no embolimoi or exairesimoi days. For refutation of this theory, see above pp. 14–16.

cutter, favoring the retention of Achaios in 166/5. Whatever the date, Meritt's calendar plays havoc with the religious and civic life of the Athenians. The universal position that Hekatombaion 1 kata theon and kat' archonta = Prytany I 1 in the fourth and following centuries is surely correct. The critical equation for Achaios' archonship is in *IG* II^2 947:

> Mounichion (X) 12 kat' archonta = Thargelion (XI) 12 kata theon = Prytany XI.12

Thus with 47± days to go in the kata theon calendar, there were 77± in the kat' archonta festival calendar. The answer surely is not Meritt's dual archonship. The authorities had to delete 30± dates in the festival calendar, sprinkling festivals among the remaining 47 days. Turning to the table of Meritt's embolimoi days, I suggest that the difference between his 20 omitted days in one archonship and 30 days in Achaios' year is only a matter of relativity. Embolimoi days require exairesimoi ones. Meritt could not propose this solution without renouncing one of his mainstay theories in treating the omitted day, i.e., that Skirophorion was sacrosanct. It is completely unconvincing that any archon's year ended Thargelion *ultimo*. I find that scholars who endorse Meritt's theories of the omitted day give no evidence that they know what theory they are endorsing.

Whereas Meritt acknowledges that the omitted 64-day calendar is an "astronomical calendar," made only for astrologoi, he nowhere throughout all of his studies charts a year omitting every 64th day, or reconciles it with his theory of alternating full and hollow months. See below chap. 4 n. 8.

2. 29th day with forward count (ἐνάτη)[6]

In *Hesperia* 4 (1935) 526–585, Meritt devoted a fascicle of his

[6]For Meritt's seven examples of forward count, see *Calendars of Athens* 26–30 and *Stone* 350 ff.

journal to a 'guaranteed' demonstration that the Athenians used forward and backward count in the last decade.[7] In this lengthy study, there is no reference to any literary source, Solon, Aristophanes, Proklos, Plutarch, etc.[8] He devoted only pages

[7]This fascicle-length article in *Hesperia* 4 525–590 was designed to show against Kirchner and a few others that backwards count was possible in the last decade. Meritt presented the text of a completely preserved stele from the Agora of the year of Ergochares (226/5) giving the text Μεταγειτνιῶνος δευτέραι μετ' εἰκάδας = Prytany III 27. *IG* II^2 838 from the same month and prytany was already known with Kirchner's text Μεταγειτνιῶνος ἐνάτ[ηι καὶ δεκάτηι δ]ευτέραι ἐμβολίμωι = Prytany III 20. Meritt, in substituting ἐνάτ[ηι μετ' εἰκάδας] for ἐνάτ[ηι καὶ δεκάτηι], made two remarkable assumptions: 1. The phrase δευτέραι ἐμβολίμωι meant one intercalated month. He wrote (p. 531), "The significance of δευτέραι ἐμβολίμωι has been correctly pointed out by Dinsmoor, and we know that we have to deal with the day ἐνάτηι μετ' εἰκάδας in an intercalated month." 2. His own equation in the new inscription is assigned to the same month, meaning that the mason omitted the embolimos phrase. Meritt puts the number of the month in brackets without telling the reader what he has done. In *Athenian Year* 58 n. 34, Meritt wrote, "Backward count with μετ' εἰκάδας has been formally proved since the publication of my article in *Hesperia* IV." The reader will find none of this in Woodhead, *Athenian Agora* XVI p. 321 in his republication of the Agora inscription. I cite this as an example of what Meritt assumes in one text (the omission of two words) he criticizes when another author assumes any omission.

[8]In *Hesperia* 4 (1935) 534, Meritt wrote that A. Mommsen had advanced the theory "with support from ancient authority." On the page cited (120), Mommsen referred to no ancient source. In his *Feste der Stadt Athen* (1898) 60 n. 4, A. Mommsen reversed his opinion: "Man kann in dem von Himerios entworfenen Bilde, wenn anders er einen Dritten vom Ende im Auge hat, den Zug vermissen, daß der Mond als feiner Silberfaden am Himmel stehe; wenn zwischen der τρίτη φθίνοντος und der ἕνη καὶ νέα, dem Tage der Konjunktion, die δευτέρα φθίν. liegt, so wird sich die schwindende Sichel an der τρίτη φθίν. zuletzt zeigen. Mitunter aber zeigt sie sich doch auch unter dieser Voraussetzung nicht, und falls in den hohlen Monaten Athens wie zu Rhodos (Newton, Greek Inscr. in the British Mus. n. CCCXLIV p. 116–119) der vorletzte Tagname, δευτέρα φθίν. (zu Rhodos Protriakas genannt) wegfiel; der τρίτη φθίν. also unmittelbar die ἕνη καὶ νέα folgte, so fiel die τρίτη φθίν. zu Athen in den hohlen Monaten

532–535 to the problem of the omitted day. On the basis of restorations, he concluded that *IG* II[2] 339 (333/2) proved that the omitted day in a hollow month was earlier than the day ἕκτη φθίνοντος (p. 533). He wrote in regard to forward count with μετ' εἰκάδας, "No proof from the inscriptions can as yet be adduced to show which day of the last decade was omitted in the hollow month." As his tables indicate, Meritt formulated the theory that ἐνάτη was always the omitted day whether with φθίνοντος or μετ' εἰκάδας or with forward or backward count. Thus in a hollow month with forward count and μετ' εἰκάδας, the omitted date was the 29th. More than twenty-five years later, after Meritt discovered the Aristophanic scholia, he still maintained in his *Athenian Year* (1961) that the count with μετ' εἰκάδας in the last decade of the month might be forward or backward. When forward count was used, he posited that the omitted date was the one before ἕνη καὶ νέα, the 29th. The following table reproduces from *Year* 58–59 his table for the count with μετ' εἰκάδας "from the end of the fourth century B.C.":

Forward Count
Full and Hollow

δευτέρα μετ' εἰκάδας
τρίτη μετ' εἰκάδας
τετρὰς μετ' εἰκάδας
πέμπτη μετ' εἰκάδας
ἕκτη μετ' εἰκάδας
ἑβδόμη μετ' εἰκάδας
ὀγδόη μετ' εἰκάδας
ἐνάτη μετ' εἰκάδας or
ἕνη καὶ νέα

---

gestrichen worden und hat Unger richtiger als ich, Chron. S. 120, geurteilt." Mommsen anticipated what I wrote in *CP* 54 (1959) 151–157.

This theory was now supported primarily on the basis of his interpretation of "elapsed days" between calendar equations where the festival and prytany calendars were not in accord, which he stated in *Year* 39, 52–53, were "guaranteed."[9] His tables show three possibilities for the interpretation of a date. For example ὀγδόη μετ' εἰκάδας might mean the 23rd, the 22nd, or the 28th; ἑβδόμη μετ' εἰκάδας might mean the 22nd, the 23rd, or the 27th. The reader of a prescript with a date μετ' εἰκάδας had to know whether the month was hollow or full, and whether the counting was backward or forward when he wanted to establish a specific date. If officials ordered an embassy to depart or troops to attack on ὀγδόη μετ' εἰκάδας, they were in a state of perplexity. And so with all civic life. Any advance notice created a dilemma. If the decision about a full or hollow month was made on the 20th day, as one recent theory has it (Walsh, *ZPE* 41 [1981] 107 ff.), no accurate advance notice could be made before that date. One wonders how parapegmata could mark such progression of days. Meritt offered no clue as to why the Athenians fashioned two lunar calendars in which the name of the day ἐνάτη was the differential feature, but it is noteworthy that for a period from 1934 to 1964 Meritt 'guaranteed' that the day before ἕνη καὶ νέα was omitted, a theory which in the latter part of his career he vigorously opposed without a backward glance.[10] As analyzed in

[9]Since Meritt gave up his theory of forward count, it follows that the method he used to establish this theory led him into error, namely the treatment of "elapsed days." However, without a backward glance, he continued to use this method almost exclusively when treating the omitted day. His examples of "elapsed days" with forward count can only be explained with backwards count on the assumption of exairesimoi and embolimoi days. If we take these as examples of what other archons could do, the "elapsed days" theory collapses. Moreover, the theory that the last day in the count (ἐνάτη = 29) is the omitted day is also abandoned without explanation.

[10]Meritt renounced forward count in a footnote (*TAPA* 95 [1964] 225 n.

Calendars of Athens 26–30 and *Ancient Athenian Calendars on Stone* (1963) 350–353, Meritt presented seven "hard-core" examples where he claimed forward count was required.[11] He assumed that the state retarded the festival calendar by as much as eight days and that the prytanies were irregular in length. Without a backwards glance, Meritt never returned to a restudy of these hard-core examples, but wrote that David Lewis had persuaded him that he was wrong.

The theory of forward count was built on the interpretation of elapsed days between calendar equations in which the count of days in the month was not the equivalent of days in the prytany. The oddity is, as we shall see, that while perforce abandoning such interpretation for these examples, Meritt continued to use the

---

86), which he later (*AJP* 95 [1974] 278–279) chided Samuel for neglecting, writing: "I have been moved to adopt this solution (of abandoning the theory of forward count) after discussion with David Lewis, and because of our knowledge now that archon's 'tampering' could omit as well as add days in the festival calendar." Meritt here endorses the theory of subtraction of days before intercalation.

[11]The first of Meritt's "hard-core" examples "which can be satisfied with nothing but forward count" (*Year* 51) occurs in his restorations for the text of *IG* II² 504. *IG* II² 505 gives the equation for the year 302/1: Skirophorion (XII) 21 = Prytany XII 21. In turn, the fragment *IG* II² 504 gives the text [Σκιροφοριῶνος δε [· · · · · ·13 · · · · · · ·, δευτέ]ραι καὶ εἰ[κοστῆι], which Meritt, following Wilhelm, restored for the following day: δε[υτέραι <μετ> εἰκάδας, δευτέ]ραι καὶ εἰ[κοστῆι]. A "simple scribal error" proved forward count. In *Calendars* 27, with backwards count I suggested that the mason misread numerals, Day 22 (ΔΔΙΙ) as Day 20 (ΔΔ), for which I cited Meritt's parallels in *Year* 112 and 161: δε[κάτει προτέραι]. If a word was omitted, as Meritt posits, in *Stone* 352 I suggested δε[κάτει <ὑστέραι> ἐμβολίωι], citing Meritt's numerous readings of epsilon iota in one letterspace. The interesting feature is that Meritt and his satellite Woodhead never retracted the text with forward count: *Hesperia* 33 (1964) 10–11; *SEG* 20.339; *Athenian Agora* XVI (1997) 195. In spite of Meritt inveighing against my restoration in *Year* 52, I tabulated in table 14 p. 374 of *Stone* the many irregularities he has assumed in his restorations in his *Athenian Year* alone.

same argument about elapsed days in maintaining his theory of the omission of day 22.

3. 22nd Day with Omission of ἐνάτη μετ' εἰκάδας in Backward Count

As observed in footnote 7 above, Meritt's initial presentation of the theory that ἐνάτη with backward count was the omitted date (Day 22) in the last decade was founded on entirely false interpretations. In publishing the text of a complete stele of the year of Ergochares (*Hesperia* 4 [1935] 525–540), he interpreted (pp. 530–581) his new equations along with those of *IG* II² 838 according to his count of "elapsed days", a method he never abandoned, to give equations completely wrong. The texts of the two inscriptions show that there were embolimoi days, and, interestingly, that the insertion of more such days continued after the date given in *IG* II² 838, a possibility always to bear in mind. Meritt later acknowledged his errors in a footnote, cited in n. 10, but never withdrew his theory about ἐνάτη, which he clearly had not proved, or ceased from regarding himself as the final arbiter in the matter of "elapsed days." His theory was initially founded on error.

In his *Athenian Year* (1961), Meritt had now discovered the two scholia of Aristophanes *Clouds* 1131 which he claimed reproduced the count in Solon's calendar, a count abandoned in 432 B.C. Indeed, these scholia may fairly be said to be the heart of the book, which he combined with his search for fragmentary equations giving "elapsed days". He introduced a game of terminology with one of restoration. One might think that, since his "proof" of the omission of Day 22 in *Hesperia* IV was admittedly erroneous, Meritt might argue according to his faith in the scholia that Day 21 was his omitted day throughout Athenian history. One may only speculate, but apparently as he went through all equations seeking counts for "elapsed days", he found equations that "proved" Day 22 was omitted. In *Year* he published two new Agora texts, one of which gave the date Day 21 (δεκάτη ὑστέρα). He claimed the following letters could be read in the prescript of a small, non-stoichedon fragment (*Year* 194 with fig. 4) without a right or left margin in which the name of the secretary is omitted and the number of the prytany and the name of the month are not preserved. The name of the archon in his text has only the last letters preserved:

αράτου ἄρχοντος ε
τῆς πρυτανεί
ει ὑστέραι, τριτ

His second non-stoichedon inscription gives the name of the secretary from the archon Pytharatos: *Year* 192/3 with fig. 3. The number of the prytany is lost and of the six preserved letters of the month (ριῶνος) one may restore Maimachterion, Anthesterion or Skirophorion.[12] The first equation Meritt restored to give the equation:

[Skirophorion] 21 = Prytany [XII] [2]3

[12]I discuss at length only the equation for Meritt's fig. 3. In *CSCA* 3 (1970) 209, I urged that Meritt's equation for the text of his fig. 4 was unproven and offered an equation with the month Maimakterion, differing by one letterspace. Meritt and Traill, *Agora* XV (1974) p. 91, responded: "This text has been given a different restoration in its opening lines by W. Kendrick Pritchett, *California Studies in Classical Antiquity*, III, 1970, p. 209. His version is of no value, for, *inter alia*, it violates the obvious stoichedon order of the first four lines. For that reason it has been omitted from the bibliography above." One need only glance at the photograph to see that the mason did not inscribe letters in chequer units. The engraving is free-hand, and one cannot even invoke a theory of syllabic division. Meritt labeled his text "nearly stoichedon 38–40" and offered the strange explanation for the restoration of Skirophorion, "The restoration made almost certain by the near-stoichedon arrangement of the text."

The second

[Skiropho]rion [XII] ἕνει καὶ [νέαι προτέραι] = Prytany [XII] 3[1]

As his invariable practice, Meritt offered no epigraphical commentary. He wrote of the first text, "The calendar equation of the text can be restored with certainty." The "elapsed days" between the two equations in the prytany calendar were 7, in the festival calendar 8. One day was omitted and it could not be Day 21, but for Meritt Day 22. He ultimately invoked two "rules". 1. Skirophorion was sacrosanct; embolimoi and exairesimoi days were not permitted. 2. By his rule of alternating full and hollow months, Skirophorion was hollow. In an article, "The Name of the Game is Restoration," *CSCA* 3 (1970), I showed (pp. 209–211) that entirely different equations were possible and with a full month. I added that more were possible, if we accepted the thesis that what Meritt had one archon do with embolimoi/exairesimoi or exairesimoi/embolimoi days, another might. On his photograph (fig. 3), I could see nothing of the alpha of the archon's name, and in *ZPE* 20 (1976) "Non Evidence for the Omitted Day," 190, I published the written statement of J. M. Camp II and Colin Edmonson "made with the concurrence of A. Boegehold and T. L. Shear Jr." that other names than Pytharatos were more likely, publishing a photograph in pl. VIIc. Meritt replied in "The Archonship of Pytharatos," *AAA* 9 (1977) 193–197, and, without apology for his earlier undotted reading, stated that he saw "the tip of a stroke, as of alpha", and offered the opinion of W. West, W. B. Dinsmoor Jr., and Mrs. Delmousou that his stroke could be interpreted as an alpha. He repeated the original photograph but then advanced the truly remarkable thesis that the mason of this non-stoichedon fragment, having neither the left or right margin, used syllabic division which requires Pytharatos.[13] In the heading,

[13]Bibliography on the debate is given in *SEG* 26 no. 93.

the stone is preserved only to show the epsilon of θεοί. I examined the stone again with John Camp and by measuring the amount of uninscribed space between the rho and the edge of the fracture argued that a normal alpha is precluded: *ZPE* 32 (1978), "A Note on Agora Inventory No. I 6731," 281–285, with another photograph in Pl. 13b with the light on the space in question. With the reading ρατος, other archons are possible: Kallistratos, Leostratos, Nikostratos, Pheidostratos, Philostratos, Phanostratos, Polystratos, Sostratos.[14] A. Geoffrey Woodhead, *Agora* XVI (1997) 274, chief scion of what G. Huxley (*AJP* 86 [1985] 303) termed the Princeton School of Epigraphy, whose credulity approaches mesmerism, is our most recent commentator, writing, "none save that of Pytharatos could be accommodated", printing the text with an alpha (dotted). I wrote to Prof. R. S. Stroud in Athens and asked him to report on the letter. He replied in a letter dated July 23, 2000, which he has given me permission to publish. I note that this was not written with me standing beside him and the report is published here in direct, not indirect, discourse: "I looked once again this morning at line 1 of Agora I 6731 and can report that there is nothing before PATOΥ that any responsible scholar would dream of reading. As was very clear the last time I examined this stone—and the several times before that—the mark, scar, blemish, etc. that BDM took to be part of a diagonal stroke of alpha (1) is not a stroke of an ancient letter (2) lies so firmly in the break that to interpret it as a stroke is hopelessly reckless and (3) it is at least 0.008 m. to the left of the P. The 'identical' sequence of A and P in TOΥAPX later in the line shows only 0.004 m. between the two letters." This long debate, of which I have given only an outline, illustrates the importance of

[14]Unfortunately, S. Tracy, in his two volumes on stonecutters, does not discuss Meritt's fig. 3 or 4 or give us the benefit of a date. The scripts of the two stones are quite different.

the alpha to Meritt's "omitted day" theory, being the one "sure" text which "proved" that Day 21 was not the omitted date, but apparently its successor. I suggest that much of the debate could have been avoided if Meritt provided epigraphical commentaries.

I turn now to what might be called Meritt's Game of Terminology. On p. 58 of the *Athenian Year*, Meritt presented the following table, here shortened, for backward count:

DAYS FROM THE 20TH TO THE 30TH OF THE MONTH IN CHANCERY STYLE FROM THE END OF THE FOURTH CENTURY B.C.

	Full	Hollow
20.	δεκάτη προτέρα	δεκάτη προτέρα
21.	δεκάτη ὑστέρα	δεκάτη ὑστέρα
22.	ἐνάτη μετ' εἰκάδας	ὀγδόη μετ' εἰκάδας
23.	ὀγδόη μετ' εἰκάδας	ἑβδόμη μετ' εἰκάδας
Etc.		Etc.

In *AJP* 95 (1974) 277, he repeated the same table with the caption, "Diagram C. Valid from Late 4th Century Onward."

There is no passage in the written literature that even hints that ἐνάτη μετ' εἰκάδας was omitted. Nor is there one iota of epigraphical evidence not involving restoration and a count of elapsed days.

Meritt made two assumptions: 1. At the end of the fourth century, the Athenians changed the count for days 20 and 21 from εἰκοστὴ and δεκάτη φθίνοντος to δεκάτη προτέρα and δεκάτη ὑστέρα. Meritt's date for the change is positively wrong, since in the sacrificial calendar of the deme Erchia, dated in the second quarter of the fourth century (Daux, *BCH* 87 [1963] 615), δεκάτει προτέραι occurs twice (A 6 and Γ 48) and δεκάτει ὑστέραι three times (Γ 2, Δ 2 and 41). Earlier, in 407/6 B.C.

[δεκάτ]ει π[ρο]τέραι occurs in line 29 of the reverse side of the Choiseul Marble.[15] 2. Since the thrust of his argument was that δεκάτη was omitted when the count with φθίνοντος was used,[16] as will be discussed at length in No. 4, he hypothesized that when the count with μετ' εἰκάδας was used, ἐνάτη, as the highest day in that count, was omitted. There is no literary evidence for any of this. The whole concept that because the Athenians used different terms to designate days of the month, particularly the 20th through the 29th, means that they counted days differently is absurd. Meritt played a game of terminology, claiming that such terms, although used simultaneously for lengthy periods, reflect different calendars.

[15]In Demosthenes 19.59, the count is given as εἰκάς, ὑστέρα δεκάτη, ἐνάτη, ὀγδόη. This negates any claim that Meton's alleged count was adopted or that the use of ὑστέρα presupposed that προτέρα must be used, or vice versa. The game of terminology is misguided. A scholiast on Aristophanes *Clouds* 1131 gives the count as εἰκάς, ὑστέρα εἰκάς. At Erythrai, the 20th day was designated as προτέρα (without δεκάτη): F. Graf, *Nordionische Kulte* (1985) 195. The εἰκάς was sacred to Apollo, being the day on which he was born: *Ety. Mag.* 298.1. For a decree of the Εἰκαδεῖς, see *IG* II^2 1258 (324/3 B.C.). Εἰκάδιος was an epithet of Apollo, see Wernicke, *RE* s.v. (1895) 50.60–66. The son of Apollo was Εἰκάδιος. For a boundary stone of the Εἰκαδεῖς, see *IG* II^2 2631. Cf. II^2 1596, line 12. For this society of "twenty-dayers", see R. Parker, *Athenian Religion* 336.

[16]In *BCH* 88 (1964) 464, I limited myself to saying that the term δεκάτη φθίνοντος had not turned up in Athenian inscriptions. I made the statement to correct that of Meritt in *Year* 39 that the term was guaranteed not only mathematically but also by the stoichedon writing in the text of the Choiseul Marble, where Meritt erred badly. In *Mnem.* 30 (1977) 235 n. 55, Meritt twists my term "inscriptions" to "language", and then adds that δεκάτη was used with the count in Eretria. This comment from Meritt seems out of place, since one of his prime tenets was that nothing but evidence specifically Athenian should be cited in considering his completely unique calendar. I, and many others, have always held that the Athenians might designate Day 21, as well as other days, by different terms.

4. Omission of Day 22 in count with φθίνοντος

On p. 45 of his *Athenian Year* (1961), Meritt presented the following shortened table for the omission of φθίνοντος:

DAYS OF THE MONTH IN CHANCERY STYLE
DOWN TO THE
LAST DECADE OF THE FOURTH CENTURY B.C.

	Full Month	Hollow Month
20.	εἰκοστή	εἰκοστή
21.	δεκάτη φθίνοντος	ἐνάτη φθίνοντος
22.	ἐνάτη φθίνοντος	ὀγδόη φθίνοντος
Etc.		

In *AJP* 95 (1974) 276–277, he changed to the following shortened tables with the count by φθίνοντος:

20. εἰκάς, εἰκοστή, εἰκάδες, δεκάτη προτέρα

Diagram A: From the Time of Solon down to 432 B.C.

	Full Month	Hollow Month
21.	δεκάτη φθίνοντος	ἐνάτη φθίνοντος
22.	ἐνάτη φθίνοντος	ὀγδόη φθίνοντος
Etc.		

Diagram B: Valid from 432 B.C. (Meton) to
Late 4th Century.

	Full Month	Hollow Month
21.	δεκάτη ὑστέρα	δεκάτη ὑστέρα
22.	ἐνάτη φθίνοντος	ὀγδόη φθίνοντος
Etc.		

Meritt's correction of his table in *Athenian Year*, which he typically never explains in print, is attributable to his confusion about Day 20. On p. 45, he wrote, "The 21st day was no longer thought of as δεκάτη φθίνοντος, but rather as δεκάτη ὑστέρα. This terminology (δεκάτη ὑστέρα) was attested in Demosthenes for the year 344/3, and appears epigraphically as early as 302/1 (I.G., II², 505)." In note 6 on p. 46, he states that when δεκάτη προτέρα was used, "δεκάτη ὑστέρα was not omitted in the hollow month." In the festival calendar (kat' archonta) the occurrence of one of the two dates does not mean that the other could not be exairesimos.

In *Year*, Meritt seemed unaware that δεκάτη προτέρα occurs in the Choiseul Marble (line 29 on p. 9 of *The Choiseul Marble* [407/6]) along with many examples of φθίνοντος. Thus, δεκάτη ὑστέρα must have been used for the 21st in the fifth century, and his theory of the omission of δεκάτη φθίνοντος in his *Year* collapsed, unless we are to assume that some omitted the 21st and others the 22nd, which would be absurd.

In the 1974 revision, Meritt avoided any chronological distinction for Day 20 in giving four different terms. Eikas is attested in Demosthenes 19.59 and other literary sources. Meritt shows no awareness that εἰκάς is attested in the earliest Athenian inscription (485 B.C.) which gives a calendar date: *IG* I³ 4.B lines 20–21; see the index in *Year* p. 262. Eikades is plural in number is used in Plutarch *Kamillos* 19.6. Eikoste is attested in that portion of the text of the Choiseul Marble (line 26) cut by the Ionic mason, which exhibits other anomalies (εἰκοστῆι τô μηνὸς ...). The Attic mason a few lines below used δεκάτη προτέρα and never adds τô μηνός.[17] Accordingly, Meritt gave up his chart in *Year*, and offered a scheme in which δεκάτη φθίνοντος was omitted only in the period from Solon to Meton. Thus, without a word of explanation, Meritt hypothesized (1) Meton changed the Athenian count of days, a theory lacking any textual support, one which completely misunderstands the role of this astrologos as attested

[17]Meritt erred badly in first reading the text as δεκάτη φθίνοντος, which even D. Lewis corrected in *IG* I³, as Meritt did later.

by many students of Meton; (2) Day 22 (ἐνάτη φθίνοντος) was omitted by Meton, for which again there is no textual support.

Meritt offered a variation about "elapsed days". Having taken from Dinsmoor the hypothesis that the Athenian year was a predetermined one of alternating full and hollow months with leap days,[18] he saw that if he had equations between Hollow Month Day 1 and Hollow Month Day 23 with a variation in the prytany date, he could postulate that Day 22 was the omitted one. In *Hesperia* 4 (1935) 532–535, he noted that *IG* II² 338 (333/2) provided a solid equation for a year with Hekatombaion (I) a full month: Metageitnion (II) 9 = Prytany I 39. In the same year, *IG* II² 339a is a stoichedon (25) text preserving eight letters of the calendar equation, as follows:

Line 6 νος ἕκτ
Line 7 ι δ

Meritt adopted Kirchner's restorations: [Μεταγε/ιτνιῶ]νος ἕκτ[ηι φθίνοντος, πέμπτηι κα]ὶ δ[εκάτηι τῆς πρυτανείας]. Meritt claimed the restoration was "inevitable", and "We must interpret ἕκτηι φθίνοντος as the 24th day of the month." The number of "elapsed days" would be fourteen. Only for one who regards square brackets as equivalent to a text on the stone are the restorations "inevitable." I have offered several restorations for the eight letters, most recently in *CSCA* 9 (1977) 188–189, some with μετ' εἰκάδας instead of φθίνοντος. Meritt isolates our two texts from others of the same year, but in *Year* 85 he published a table showing that there was "tampering" in this same year. Applying the same principle to our equations in the first part of the year, I proposed for the equation of II³ 339: ἕκτ[ηι φθίνοντος

[18]As we shall see, Dinsmoor claimed that Meton changed a true lunar calendar, as, for example, at the time of the laying out of the Parthenon, to one with alternating full and hollow months with leap days every 15 and 17 months.

ἑβδόμηι κα]ὶ δε[κάτηι]. No day was omitted. Since there are hundreds of examples of vacats in Greek inscriptions, we may read ἕκτ[ηι μετ' εἰκάδας, ἕv|κτηι καὶ δεκάτηι], placing the vacat where we please in the long lacuna. This restoration was so important to Meritt with his sparse collection of three examples that he repeatedly objected to this vacat as breaking the unity of the phrase. As a matter of methodology, a word about vacats is in order.

In *Athenian Year* 117, Meritt wrote: "The occasional uninscribed space in an inscription may be quite inexplicable on any attempt at rationalization. Why, for example, were six spaces left uninscribed in the phrase ἐπὶ τῆς Οἰνεῖδος vv[vvvv] ἑ[β]δόμης π[ρυτα]νείας of *I.G.* II^2 910?" A. S. Henry, *Mnem.* Suppl. 49 (1977) 63–70, discusses at length uninscribed spaces in prescripts "for the purpose of emphasis." Equally important in the game of restoration are examples which have nothing to do with emphasis. Vacats occurring in the middle of words and at the end of lines are not a matter of emphasis or punctuation. Meritt (*Hesperia* 43 [1974] 465) restores [ἕκτης vv πρυταν] in the middle of a line without commentary. In *Year* 113, the restoration is [τρίτης v πρυτανε]. In a number of texts where the margins are not preserved, Meritt's restorations have vacats at the end of lines; see the collection in *CSCA* 9 (1977) 189.[19]

For the omissions of words, Henry, in his study of prescripts, finds omissions in the lunar dates on pp. 38, 56, 72, 84; in the prytany dates on pp. 38, 56, 60, 62, 67, 72, 85, 88.

[19]For numerous vacats in Meritt's own texts in his *Athenian Year*, see pp. 86, 94, 96, 105, 113, 119, 120, 126, 127, 128, 143, 157, 164 (bis), 170, 179, 198, etc. S. V. Tracy, *Hesperia* 69 (2000) 227–233, collects examples of blank spaces to emphasize the names of orators in decrees 307/6–302/1.

In *Stone* (*UCPCA*) 1963 310, I published tables of textual anomalies in the carefully inscribed Logistai inscription (*IG* I^2 324), one showing anomalies on the stone, the other anomalies in Meritt's restored text. Another type of irregularity assumed by Meritt in endorsing M. Lang's text of the same inscription is that there were imperfections in the marble in the missing section which resulted in vacats.

If Meritt can postulate two vacats in the middle of a line and five vacats at the end of lines (*AE* 1973 241 = fig. 4 of *Year* following p. 194), and that without commentary, one may allow that other stone-cutters may have done the same. In *Calendars of Athens* 48, noting that there are five inscriptions dated within five years of II^2 339 omitting the prytany date, and that there were malformed alphas in the form of deltas (n. 26), we suggested the possibility of the reading ἕκτ[ηι ἐπὶ δέκα· ἐκκλησία κυρ]ί<α> ἐ[ν ... . Meritt confirmed that the letter was formed as an alpha, but artfully did not address the problem of malformed letters. This small fragment, *IG* II^2 339, affords no evidence for the omission of Day 22. In the year following our inscription (i.e. 332/1), we have a beautiful inscription with carefully inscribed letters labeled stoichedon; see the photograph in *Hesperia* 8 (1939) 26–27, No. 6. Woodhead, *Agora* XVI (1997) 121, describes the lettering as follows:

> The writing is "standard" for its time, with slight emphasis of "free ends" and some carelessness, for example, in the inscribing of epsilon or the alignment of a stoichos. In line 9 three letters are set in the space of two, and in line 7 two are restored in the space of one. At line 19 an entire line may have been omitted.

Thus, in one carefully inscribed "stoichedon" inscription, we have vacats, omissions, and multiple letters in spaces. We need a handbook which would tell the student something about the irregularities of stonecutters in 'stoichedon' and non-stoichedon texts.

The one and only set of equations preserved sans restoration where Meritt claimed by his "elapsed days" theory that ἐνάτη μετ' εἰκάδας was omitted is set forth in *AE* 1968. 78–80 for Skirophorion in the intercalary year 303/2 B.C.

In *IG* II[2] 493, 494, we have the equation:

Σκιροφοριῶνος (XII) δεκάτει ὑστέραι, τρίτει καὶ εἰκοστεῖ τῆς πρυτανείας

In *IG* II[2] 495, 496, 497:

Σκιροφοριῶνος ἕνηι καὶ νέαι προτέραι, μιᾶι καὶ τριακοστῆι τ. π.

Meritt wrote, "No ingenuity on the part of an unbeliever can manipulate these dates by the substitution of alternative texts," and continued that they proved that the month must be hollow with ἐνάτη μετ' εἰκάδας the omitted date, offering the following table:

Skirophorion 21	=	δεκάτη ὑστέρα
Skirophorion 22	=	ὀγδόη μετ' εἰκάδας
(ἐνάτη μετ' εἰκάδας omitted in this hollow month)		
Skirophorion 23	=	ἑβδόμη μετ' εἰκάδας
Skirophorion 24	=	ἕκτη μετ' εἰκάδας
Skirophorion 25	=	πέμπτη μετ' εἰκάδας
Skirophorion 26	=	τετρὰς μετ' εἰκάδας
Skirophorion 27	=	τρίτη μετ' εἰκάδας
Skirophorion 28	=	δευτέρα μετ' εἰκάδας
Skirophorion 29	=	ἕνη καὶ νέα προτέρα
Skirophorion 30	=	ἕνη καὶ νέα (ἐμβόλιμος)

Meritt wrote (p. 80) that his table "shows that δευτέρα μετ' εἰκάδας was not available to them (Athenians) for the 29th. It had, in fact, already been used for the 28th," and one wonders why a scholar so experienced with the calendar tries to pass off such an error.[20] Meritt's table and his conclusions are erroneous. If a single date is taken out of the calendar, the term for that date was omitted and on the parapegma the peg for that day was skipped. If an exairesimos day was taken out between δεκάτη ὑστέραι and ὀγδόη μετ' εἰκάδας, the term applied to the day after δεκάτη ὑστέραι was ὀγδόη μετ' εἰκάδας, but it was Day 23 in the calendar or on the parapegma. The day ὀγδόη μετ' εἰκάδας on the calendar or parapegma was never Day 22, but Day 23. All exairesimoi dates were days taken out of the calendar. Conversely if the day τρίτη ἱσταμένου δευτέρα ἐμβόλιμος, for example, was given, the count was 3, 3^1, 3^2, with the next day being τετάρτη ἱσταμένου, although the 6th day of the month. Again, when Meritt (*Hesperia* 32 [1963] 437) posits only 17 days in a month with 12 being exairesimoi, it does not mean that the last day of the month was termed ἑβδόμη ἐπὶ δέκα. The last day, if it was not exairesimos, was always ἕνη καὶ νέα.

The days between our two equations proceeded in regular fashion:

Prytany date		Festival date
23	=	δεκάτη ὑστέρα (21)
24	=	ἐνάτη μετ' εἰκάδας (22)
25	=	ὀγδόη μετ' εἰκάδας (23)
26	=	ἑβδόμη μετ' εἰκάδας (24)
27	=	ἕκτη μετ' εἰκάδας (25)

[20]Playing his game of terminology, Meritt must have known what he was doing. In *ZPE* 41 (1981) 123 n. 27, J. A. Walsh writes, "Owing to the necessity of restoration, some of the epigraphical evidence has been mooted, but not the evidence of 303/2, ... to which critics of δεκάτη/ἐνάτη φθίνοντος being the omitted date have not adequately responded." Walsh's paper demonstrates such an ignorance of the festival calendar and its literature that one would think he would want it mooted. If a day had been taken out of the calendar between Day 21 and Day 30, it is obvious that it could be δευτέρα μετ' εἰκάδας.

Prytany date		Festival date
28	=	πέμπτη μετ' εἰκάδας (26)
29	=	τετρὰς μετ' εἰκάδας (27)
30	=	τρίτη μετ' εἰκάδας (28)
31	=	δευτέρα μετ' εἰκάδας (29)
32	=	ἕνη καὶ νέα προτέρα (30^1)
33	=	ἕνη καὶ νέα(30^2)

There was no omitted day and the month was full in a year of 384+1 days, normal for a lunar calendar; see the table in *Calendars of Athens* 69. The word προτέρα was applied only to Skirophorion ἕνη καὶ νέα at the end of the year, never to ἐνάτη μετ' εἰκάδας, a strong clue to an observation calendar with particular attention at the end of the year. For the term proteros, see *Ancient Calendars on Stone* 349: *IG* II² 486, 495, 496, 497, 597 add., *Year* 192–193, *Hesperia* 7 (1938) 297.[21]

The *AE* 1968. 78–80 equations constitute Meritt's sole example with uncontested texts to sustain his theory of the missing ἐνάτη by his count of "elapsed days". It seems odd that a scholar with his knowledge of the calendar would so misrepresent the count, making the claim that δευτέρα μετ' εἰκάδας "had, in fact, already been used for the 28th", and so could not be the omitted day. In contrast with his other two cases, one with a misread text, the other with a five-letter equation at best to be offered

[21]Out of context but in his commentary on *IG* II² 493–497, Meritt injects the following note (p. 78 n. 4): "Pritchett's attempt to throw out the Aristophanic scholia altogether because there was, as he claims, no δεκάτη φθίνοντος is valid only if he can prove that there was in fact no day so called. It will take a lot of proving." This is an obvious untruth. After correcting Meritt's text of δεκάτει φθίνοντος in line 29 of the Choiseul Marble to [δεκάτ]ει π[ρο]τέραι, I merely observed that δεκάτη φθίνοντος had not yet turned up at the time in Athenian epigraphical texts. That the grammarians, scholiasts, and Athenians used it, I have never questioned.

exempli gratia, the equations in *IG* II^2 493–497 provided the only example by Meritt of epigraphical texts sans restorations which he cited for his ἐνάτη theory. All three claims are unproven. The bizarre theory that calendars with φθίνοντος and μετ' εἰκάδας omitted different dates is entirely Meritt's invention.

Conclusion. The thrust of this chapter is to show that there is no epigraphical evidence supporting the theory of the omission of Days 21/22. Of the "hard-core" and "guaranteed" examples, one involves a false reading of a small non-stoichedon fragment, another a restoration of even a smaller fragment permitting many restorations, and the third a misunderstanding of exairesimoi days. We may also lay to rest any theory that Skirophorion was sacrosanct, that any year ended on Thargelion (XI) ultimo, that a calendar omitting every 64th day regulated the festival one, that Hekatombaion 1 did not equal Prytany I 1, that any year began with Skirophorion, or that a count of "elapsed days" provides firm evidence in a calendar often exhibiting exairesimoi days.

In contrast with Days 21/22, there is strong epigraphical evidence supporting the hypothesis of the omission of δευτέρα μετ' εἰκάδας. Part of the evidence is in the fact, noted by Daux and others, that in the Erchia (Daux, *BCH* 87 [1963] 603–634) and four other sacrificial calendars of the demes (*SEG* 26.138), no sacrifice or festival is prescribed for Day 29 in contrast with those on Days 21 and 22. In L. Deubner, *Attische Feste*, his fold-out table also shows no festival on Day 29. Of equal significance is the fact that in Jon D. Mikalson's *Sacred and Civil Calendar of the Athenian Year* (1975), meetings of the Boule or Ekklesia on Day 29 are attested for every month but Hekatombaion. As for Hekatombaion, A. Mommsen, *Feste* (1898) 153, using the catalogue of prizes listed in the inscription now published as *IG* II^2 2311, estimated that the Greater Panathenaia ran for seven days, Days 21–29, giving three days to musical contests. Parker, *OCD*^{3}

1104, suggests that the major athletic and musical contests applied only to the Greater Panathenaia, open to all Greece and held every four years. The M scholia to Euripides *Hekuba* 467 says the Panathenaia lasted four days. The B scholiast to Aristides *Panathenaikos* 147.14 (Dindorf) writes τρεῖς ἡμέρας ἤγετο τὰ Παναθήναια, whereas the C scholiast says ἤγουν αἱ τέσσαρες ἡμέραι. The lack of any evidence for Day 29 as a festival or sacrificial day, in contrast with Days 21 and 22, suggests that Day 29 was avoided because it was the omitted day in a hollow month.

III

LITERARY EVIDENCE FOR OMITTED DAY

The literary evidence for the omitted day overlaps with that for the lunar nature of the festival calendar, but is here treated separately, since Meritt and his followers focused chiefly on this feature. While Meritt advanced the theory in his *Athenian Year* (1961), held by no earlier student of the calendar including himself, that the two scholia in Aristophanes *Clouds* 1131 provided reliable evidence for the omission of Day 21, his major effort subsequently has been to downplay the contradictory, and more substantial, testimony that Day 29 was omitted. There is positively no literary evidence for the omission of Day 22, and, as has been seen in Chapter II, no epigraphical. Meritt, in his game of terminology, held that only in the period between Solon and Meton (432) was Day 21 omitted.

Turning to the literary evidence, I first treat the problem of whether the scholiasts on *Clouds* 1131 in the tenth and later centuries of our era could have known such a detail about Solon's calendar, allegedly abandoned in 432 because of a change in terminology. The literary evidence is treated under the following headings:

1. Solon
2. Scholia to Aristophanes *Clouds* 1131
3. Pollux
4. Proklos
5. Moschopoulos
6. Geminos

1. Solon. Meritt claimed throughout his studies that Solon's calendar "is the calendar described by the grammarians and the

scholiasts", for example, *AJP* 95 (1974) 276 n. 27. In *ZPE* 41 (1981) 113, J. A. Walsh wrote, "It is to the credit of Solon's genius that he was able to continue the use of a lunar calendar while modifying it in certain important respects. Under Solon a decision on the month's length was reached before the month began—perhaps according to the paradigm of a cycle. This decision was checked by observation of the moon's phases until the beginning of backward count", when he claims Days 21 and 22 were omitted.

Our knowledge of Solon is not extensive. What we know of his calendar comes largely from Plutarch. One authority, Ivan Linforth, *Solon the Athenian* (Berkeley 1919) 296, has summarized the evidence as follows:

> Plutarch attributes to Solon certain changes in the Athenian calendar. "Observing," he says (*Sol.* XXV, Perrin's translation), "the irregularity of the month, and that the motion of the moon does not always coincide with the rising and setting of the sun, but that often she overtakes and passes the sun on the same day, he ordered that day to be called the Old and New, assigning the portion of it which preceded the conjunction to the expiring month, and the remaining portion to the month that was just beginning. ... After the twentieth he did not count the days by adding them to twenty, but by subtracting them from thirty, on a descending scale, like the waning of the moon." This is obviously an invention to explain the two features of the calendar which are mentioned. Such things as these would not have found a place in any record, and there is no way, so far as we can see, by which Plutarch or his sources could have actually known that Solon made such innovations. The peculiarities, for which an origin was sought, were more likely the result of popular habit. The attribution of them to Solon may rest entirely on the fooling of Pheidippides in the *Clouds* of Aristophanes (1187 ff.), who claims that the ἕνη καὶ νέα was devised by Solon as a popular measure to provide a respite for men who were threatened with a lawsuit.

Plutarch, in using such terms as σύνοδος, φθίνοντος, ἱσταμένου, ὥσπερ τὰ φῶτα τῆς σελήνης, is clearly referring to a calendar regulated by the moon, not a schematic one of alternating full and hollow months. Dover, in his commentary on the *Clouds* 1187, observes that just as we speak of English law or French law, the Greeks spoke of the laws of Solon, and Andokides (1.95) actually describes as 'the law of Solon' an enactment which he quotes and dates to the summer of 410 B.C.[1] Pheidippides (*Clouds* 1179 ff.) is quibbling in arguing that the ἕνη καὶ νέα means two different days. Any calendar with ἕνη καὶ νέα might be called that of Solon.

E. Ruschenbusch, *Solonos Nomoi* (*Historia* Einzel. 9 [1966]) "Opferkalender," frgs. 83–86, pp. 101–102, collects passages which include references to θυσία ἐν τοῖς ἄξοσιν. In the collection of Fasti Sacri in *IG* I³, we have one inscription (I³ 234, dated 480–460) which itemizes sacrifices to certain deities. Lines 15–16 read, Γαμελιôνος: με[νὸς - - - φθί]νοντος. L. H. Jeffery, "Boustrophedon Inscriptions from the Agora," *Hesperia* 17 (1948) 86–111 with pls. 29–32, collects a number of fragments, some relating to sacrifices at the Mysteries (*IG* I³ 231). Solon may have prescribed the nature of offerings and prices for certain festivals, possibly giving a specific date, but axones were concerned with religious festivals, and there is no suggestion that they could be so detailed as to give in consecutive order the count of days throughout any month. To future researchers, whether grammarians or schoolmasters, the axones could have given no information about full and hollow months or an omitted day. Meritt and his followers even object to drawing any conclusions from the detailed Erchia calendar. Jeffery comments on the difficulty of assessing the true Solonic matter amidst the mass of

[1]Cf. Andrewes, *CAH*² III.3 (1982) 376, "irresponsibly attaching the name of Solon to laws manifestly later."

material. R. Parker, *Athenian Religion* (1996) 43–55, gives up-to-date bibliography.

We can be sure that no Byzantine scholiast or grammarian knew anything of a calendar peculiar to the period between Solon and Meton. And if they had, they would not have presented it as the only calendar ever used by the Athenians.

2. Scholia to Aristophanes *Clouds* 1131. Meritt, throughout several studies, claimed that the scholia of the *Clouds* gives the count of days according to Solon.[2] We have a host of scholia on Aristophanes' statement when Strepsiades enters counting the days. The reader is urged to consult the fold-out table facing p. 44 in W. I. W. Koster, *Scholia in Aristophanis Plutum et Nubes* (1927). Koster lists in ten columns different tables for the count of days classified, not according to manuscripts, but terminology. In the multi-volume Groningen publication *Scholia ad Aristophanem*, edited by Koster, we have many lists collected from many manuscripts, sometimes more than one in a manuscript, as with the Ravennas and Venetus, as follows:

Vol. I.3.1	=	Pages 210–211
Vol. I.3.2	=	Pages 158–160
Vol. 4.2	=	Pages 637–640

Although giving only the scholia in Codex Ravennas, W. G. Rutherford, *Scholia Aristophanica* 1 (London 1896) 248–250, reproduces in the middle tier on each page the exact Greek text as it appears with abbreviations and ligatures, which the reader might consult. The third tier is translation. By contrast, in the Groningen

[2]N. G. Wilson, *Scholars of Byzantium* (London 1983) 33–36, discusses the pros and cons of the date of the introduction of commentaries on ancient texts. The codex, instead of the roll, made it easier to give explanatory material. This type of book became standard for all types of texts by the end of the fourth century.

publication of Koster (*Scholia ad Aristophanem* I.3.1.210 ff.), the text is spelled out with the abbreviations and ligatures in the notes. The writer of the scholia of the Ravennas and Venetus recorded two lists, one with δεκάτη φθίνοντος as the 21st day, the other with ἐνάτη φθίνοντος as the same day. For example, the Venetus reads: ἕως εἰκάδος· εἶτα ἣν ἡμεῖς κα ἐνάτ(ην) αὐτοί φα(σιν). Since there is a difference of a day, the two tables have been invoked as one giving the table for a full month, the other for a hollow month with δεκάτη φθίνοντος being the omitted day. None of the tables refers to a full or hollow month. They merely list the terminology for the count of days in the Attic month in contrast with what ἡμεῖς (φαμέν). There are many ligatures. There is no list which reproduces exactly what is found in Attic chancery style of any period.[3]

There is one firm piece of evidence that the scholia to the *Clouds* do not represent the count of days in chancery style in the period between Solon and Meton. In one of the earliest Athenian inscriptions which gives dates in terms of months and days, Day 15 is twice given as διχομηνία, a word not found in any of the scholia. The word is so used for this day in the calendars of other states; see A. E. Samuel, *GRC* p. 103 for Tenos, 110 for Rhodes, 113 for Cos. For Priene, see Hiller, *Inschriften von Priene* no. 4 line 45. In Plutarch *Dio* 23.3, the text reads ἡ δὲ σελήνη διχομηνίαν ἦγε. The Souda has διχομηνιαία πεντεκαιδεκαταία τοῦ μηνὸς ἢ τῆς σελήνης. The text is *IG* I³ 6.B, dated c. 460, which gives the durations of the sacred truces, first for the Greater Eleusinian Mysteries, then for the Lesser Mysteries. The former was to run for c. 55 days from the διχομηνία of Metageitnion

[3]One of the lists confirms the statement of Plutarch (*Demetrios* 12.1) in attributing the term Δημητριάς to the ἕνη καὶ νέα. The term μεγάλη εἰκάς for the Athenian 20th day in several of the lists is puzzling. Perhaps the term derives from the fact that Boedromion 20 was an important day in the celebration of the Mysteries, when Athens was flooded with visitors.

through Boedromion up to the δεκάτη ἱσταμένου of Pyanopsion (lines 19–27); the latter from the διχομηνία of Gamelion through the Anthesterion to the δεκάτη ἱσταμένου of Elaphebolion (lines 33–47). The truce was to be good in the various cities that sent worshippers to the Eleusinia, as well as at Athens, for those who take part in the festivals (lit. 'use the temple'). See Latte, *RE* s.v. Spondophoroi (1929) 1949.26–30. The terminology is that of a lunar calendar, understood in other states.

An even earlier Athenian inscription also testifies to a well-organized lunar calendar. In the famous Hekatompedon inscription of 485/4 (?), *IG* I³ 4.B, lines 18–21 read: "the treasurers are to open and inspect the houses on the same days, always on the first of the month, on the tenth of the month, and on the twentieth of the month", νομενία, δεκάτε, εἰκάς.[4] The intervals indicate every ten days.

The scholiasts in copying their sources reproduce the terms ἡμεῖς (φαμέν), αὐτοί φασιν, ἕνη καὶ νέα οὕτως παρ' Ἀθηναίοις ἡ παρ' ἡμῖν τριακάς, etc. What is the date of these subjects in the present tense? It would be utterly absurd to claim that they were contemporaries of Solon or of Meton. The references are clearly to contemporary Athenian and Greek calendars. The scholiasts were copying, through whatever chain of transmission one may hypothesize, texts which characterized a contemporary calendar (ἡμεῖς [φαμέν], παρ' ἡμῖν).

W. J. M. Starkie, who offered a study of the scholia of Aristophanes, concluded in his commentary on lines 1131 ff., after citing R and V:

> These notes are hideously verbose, and they omit the one fact worth knowing, viz. that in the 'hollow months' (of twenty-nine

[4]I give B. Jordan's translation of the text offered exempli gratia by R. Renehan: *Servants of the Gods* (= *Hypomnemata* 55. [1979]) p. 49. For the usual text and translation, which seems to me faulty, see Jordan's p. 21.

days) δευτέρα φθίνοντος was left out (ἐξαιρέσιμος), and the twenty-ninth day was called τριακάς or ἕνη καὶ νέα; see G. F. Unger in I. Müller, *Handb.* i. pp. 564 § 13, 565.

On pp. lxiii–lxxi, he points out that in contrast to the *Aves* where the scholia are excellent because the play was not familiar to schoolmasters, it was different with sections of the *Clouds* and *Plutus* which were popular in schools and attracted a lot of rubbish which passed under the name of scholastic interpretation.[5] He writes (p. lxv):

> In this way an immense edifice of comment must have been erected in connexion with certain familiar texts. The 'anaemic' and prolix commentary on the *Plutus* and *Clouds* is probably a compilation of σχολικὰ ὑπομνήματα of this kind, by generations of school-teachers whose names are deservedly forgotten. It is significant of the provenance of such lore, 'chewed by blind scholiasts o'er and o'er,' that the names of the great Alexandrine scholars are conspicuous therein by their absence.[6]

Starkie, who devoted Chapter IX of his Introduction to the *Clouds* to the scholia, also wrote in part (p. lxiii):

> *The Origin of the Corpus of Old Scholia*
>
> The only direct evidence available as to the provenance of the corpus of old scholia is contained in the 'subscriptions' to the *Clouds*, *Peace*, and *Birds*, which are to the following effect: 'The *cola* have been divided according to <the manuscripts> of Heliodorus: the marginal commentary has been compiled from the scholia of Phaeinus and Symmachus and some others'; 'the *cola* have been divided with reference to <the manuscripts> of Heliodorus:

[5]There are over a hundred and thirty medieval manuscripts of the *Clouds*. Both in antiquity and in the Middle Ages it was a popular comedy.

[6]For errors in the old scholia of the *Clouds*, see *Ancient Athenian Calendars on Stone* 325 n. 48.

> the marginal commentary has been compiled from the notes of Phaeinus and Symmachus'; 'the marginal commentary has been compiled from the notes of Symmachus and other scholia'. All scholars agree to decry Phaeinus, whose name occurs five times in the commentary on the *Knights*, as an ignorant Byzantine; while, since the essay of O. Schneider 'De ueteribus in Aristophanem scholiorum fontibus,' Symmachus has generally been considered to be the author of the corpus of old scholia, and indeed their 'sole fountain.' But even Symmachus, who flourished about 100 A.D., cannot be regarded as an original writer. Long before his time, the golden age of 'the heroes' of criticism, to use Rutherford's phrase, had come to an end, and the scholiasts were content to reproduce, as their own thoughts, the notes of the older men derived sometimes from their commentaries, sometimes from the lecture-notes of their pupils.

K. Dover, in his summary of work on the scholia of Aristophanes in *Lustrum* 2 (1957) 107, writes:

> Zuntz (221) in the first part of his article edits the Ar. papyri which contain marginal comments, distinguishes, on generic grounds, the source or type of source of each element in those comments, and exhibits their relationship to the scholia of our mss. In the second part he argues that the 'corpus' of scholia vetera does not, as a corpus, antedate s. IX; White's supposition that it was formed in antiquity seems to be invalidated by the late date (s. VIII) at which Christian texts were first furnished with marginalia which amounted to commentary. When the corpus was compiled in s. IX, it was not a question of transferring the contents of a single hypomnema en bloc to the margins of a codex, but of collecting the comments found in varying quantity in the margins of texts which had survived from antiquity.

D. Holwerda, in his commentary on one of the $\overline{\kappa\alpha}$ = ι΄ φθίνοντος lists in Koster's Vol. 4.2 p. 638, wrote: "Ego post κα΄ nonnulla excidisse coniecerim collato scholio, quod praebet E (non apud Db), e.g. ἣν ἡμεῖς κα´ <μίαν καὶ εἰκάδα ἣ ι´ φθίνοντος, ἣν δὲ ἡμεῖς κβ´> θ´ φθίνοντος κτλ.”

If the scholiasts of MSS., none earlier than the tenth century of our era, were copying two lists one for a hollow month and the other for a full one, they failed to give any clue. All lists contain alphabetical numerals, which were easily corrupted. More importantly, it is easy to skip a term in the count. I follow Boeckh, Unger, Starkie, Holwerda, and others in discrediting the table which omits the day for ι´, however it is to be explained.[7] I favor the solution of Holwerda. The scholiasts of R and V derive from some source which omitted a line. The date of those who wrote ἡμεῖς, αὐτοί φασιν is clearly not in the time of Solon or Meton, but surely many centuries later.[8] They offer no evidence for a calendar restricted to 594–432 B.C. Moreover, they give day 20 as εἰκάς, which demolishes Meritt's theory that Meton decreed δεκάτη προτέρα and δεκάτη ὑστέρα for days 20 and 21. For those who play Meritt's game of terminology, the lists give no support for the omission of Day 22, critical for the entire period where we have calendar equations. Meritt repeatedly tells his reader, “The Aristophanic scholia show that the omitted day in a hollow month came where the backward count began”, e.g. *Arch. Eph.* 1968. 79. The scholia say no such thing. It is unfortunate that such untruths are picked up by such scholars as Woodhead and Walsh and made the basis of their calendar studies.

I have treated at such length the question of the validity of the scholia to this one line of the *Clouds* because it is, in my judgment, the only passage in the literature which can be taken to contradict the theory of a true lunar calendar, e.g., the equation

[7]For alphabetic numerals, see Leslie Threatte, *The Grammar of Attic Inscriptions* 1 (1980) 113–119.

[8]Koster noted that one of the lists compares Athenian and Julian dates.

κατὰ θεόν = κατὰ σελήνην = a lunar calendar, an equation which Meritt and his followers have failed to refute in spite of hundreds of pages of writing. The theory will be addressed in Chapter IV.

3. Pollux. Meritt argued that another table of days of the month, one in Pollux 1.63, also gives the count for a hollow month in the time of Solon, omitting δεκάτη φθίνοντος. He devotes an article to the subject in *Arch. Eph.* 1973. 237–245. Indeed, he went to the trouble to publish a photograph of a page of Parisinus 2647, which Bethe labeled B, dating it to "saec. XIII." Meritt says that the following text, found in B, was "substantially the same" as Bethe's 1900 text, reprinted in 1967, except for two small variants.

The critical part of B reads:

> καὶ μέχρι τῆς εἰκοσάδος· τὸ δὲ ἀπὸ τούτου πρώτη ἐπὶ εἰκάδι· ἡ δ' αὐτὴ καὶ ἐννάτη φθίνοντος· ἐννέα γὰρ λοιπὰ ἀπὸ τῆς κα´· καὶ ὁμοίως ἄχρι τῆς τριακάδος, ἣν οἱ ἀττικοὶ καλοῦσιν, ἔννην καὶ νέαν·

Whatever Meritt meant by "substantially", he has clearly misled the unwary reader. When we turn to Bethe's text of Pollux (Vol. I pp. 19–20), we find a completely different text (omitting brackets):

> μέχρι τῆς εἰκάδος, τὸ δὲ ἀπὸ τούτου πρώτη ἐπὶ εἰκάδι—ἡ δ' αὐτὴ καὶ δεκάτη φθίνοντος—καὶ δευτέρα ἐπὶ εἰκάδι—ἡ δ' αὐτὴ καὶ ἐνάτη φθίνοντος—καὶ ὁμοίως ἄχρι τῆς τριακάδος, ἣν οἱ 'Αττικοὶ καλοῦσιν ἕνην καὶ νέαν,

In his critical apparatus, Bethe explains that M, the earliest (saec. X) manuscript, omitted καὶ δευτέρα ---- φθίνοντος. In turn, on δεκάτη he notes that B (only) had the variant ἐννάτη. Bethe continues by saying that B added φθίν. ἐννέα γὰρ λοιπὰ ἀπὸ τῆς κα´ καὶ ὁμοίως. The picture is clear. Only B read ἐνάτη instead of δεκάτη. Moreover, when B writes ἐννέα λοιπὰ ἀπὸ τῆς κα΄, it is equally clear that δεκάτη (or its ligature) must have stood in the archetype, for κα΄ (21) plus ἐννέα (9) equals 30. Indeed, this scribe called the 30th τριακάς. All the tables in various MSS. were for 30-day months, as I suggest was the case for those in the Aristophanic scholia.[9] It is inexcusable that Meritt misled the reader so about Bethe's text and does not recognize that 21 + 9 = 30. Bethe, in his Praefatio, informs us that the *Omastikon* of Pollux perished, and the extant MSS. are derived from four families of incomplete copies descending from an early epitome possessed and interpolated by Aristeas ca. A.D. 900. These four groups of MSS. are charted on Bethe's p. xx. There are errors and corruptions in all MSS. Bethe writes of Books I–VII, "Tradiderunt praeter M tria exempla: II (= F + S) et A (V) et CB," plus the Aldine. C, which has δεκάτη, is of the same family as B. It is difficult to understand how any experienced scholar would cite only B, feigning ignorance of the art of textual criticism. In fact, Meritt was responding to my treatment of the text in *CSCA* 3 (1970) 205–209, expanded later in *CSCA* 9 (1977) 184–185; so was well aware that he was misrepresenting Bethe's text established from four families of manuscripts. It is impossible to take seriously such scholarship. Moreover, Meritt represents Pollux as reproducing Solon's calendar.

[9] F. K. Ginzel, *Handbuch der mathematischen und technischen Chronologie* II (1911) 317, citing many examples of 30-day months and 360-day years, concluded, "Es ist aber kein Zweifel, daß alle diese Beispiele für nichts anderes zeugen, als für den Sprachgebrauch, welcher die runde Zahl von 30 Tagen als Monatsdauer annahm. Die 30tätigen Monate waren also ein populärer, die vollen und hohlen Monate ein kalendarischen Begriff." *Webster's New International Dictionary* notes a similar use in English: "In popular use, a period of four weeks or of thirty days is often called a month."

3. Proklos. There are two passages of Proklos which say that the Athenian omitted day was the 29th. These are treated separately.

A. Proklos on Hesiod *Op.* 765–768: ἄρχεται οὖν ὁ Ἡσίοδος ἐκ τῆς τριακάδος, καθ' ἣν ἡ ἀληθής ἐστι σύνοδος, ὁτὲ μὲν οὖσαν τριακάδα ἄνευ ἐξαιρέσεως, ὁτὲ δὲ εἰκοστὴν ἐνάτην, ὅτε καὶ ὑπεξαιρεῖται ἡ πρὸ αὐτῆς ὑπὸ 'Αθηναίων.

B. Proklos on Hesiod *Op.* 817–818: τοῦτ' ἐναργὲς ἐποίησεν, ὅτι τρίτην εἰνάδα κέκληκεν οὐ κατὰ 'Αθηναίους τὴν δευτέραν εἰκοστὴν ἀνάπαλιν ἀριθμοῦντας τᾶς φθινούσας — δεκάτην, ἐνάτην, ὀγδόην καὶ ἑξῆς —, ἀλλὰ τὴν πρὸ τριακάδος· περὶ γὰρ ταύτης ἀμφιβάλλουσιν, εἴτε ἐσχάτη ἐστίν, εἴτε πρὸ τῆς ἐσχάτης·

First a word about the source used by Proklos (A.D. 412–485).

Source of Proklos. In collecting the fragments of Philochoros, Jacoby, *FGrH* 3 B no. 328, attributes frg. 213 (p. 157) to Proklos. In this fragment, Philochoros is cited by name. More importantly, Philochoros is cited by name in frgs. 88, 189, and 190, all three assigned to Proklos. In frg. 88 the citation is Φιλόχορος δὲ ἐν τῶι Περὶ ἡμερῶν, and all three fragments are associated with the Περὶ ἡμερῶν by Jacoby. Moreover, we know from frgs. 769–771 of Pertusi's collection of the Hesiod scholia that Proklos elsewhere referred to the Athenian calendar; cf. Jacoby, 3 B Suppl. 1 p. 372, where, incidentally, Jacoby says Proklos is a better witness than the scholia to Aristophanes. These passages, completely neglected by Meritt and his followers,[10] are overwhelming evidence

[10]David Lewis, who endorsed all of Meritt's views about the calendar including those which Meritt later renounced, such as Geminos 8.52 as the authority for the Athenian calendar, wrote (*JHS* 83 [1963] 195), "M., by using the Aristophanes Scholia rather than a vague statement by Proclus,

that Proklos used an excellent authority on Athenian calendric matters.

One may note that in passing and without argumentation, Jacoby, 3 B Suppl. 1 p. 366, would have Proklos using Plutarch's Περὶ ἡμερῶν, in other words Philochoros indirectly. Jacoby writes:

> **(85–88)** The fragments of the special work Περὶ ἡμερῶν, which comprised at least two books, have been preserved in Proklos' commentary on Hesiod's *Erga* (F 88; 189–190) and in Photios' *Lexicon* (F 85–87). The former certainly took them from Plutarch's commentary, the latter from the Synagoge, which on its part probably has them ultimately from the Atticist Pausanias. The material may be increased from the Scholia on Hesiod with some certainty, for, so far as we can see, Ph. alone among the earlier authors wrote Περὶ ἡμερῶν; the author quoted in F 190 besides Ph. (certainly already by Plutarch) in regard to the days which τὰ πάτρια τῶν 'Αθηναίων καθαρμοῖς ἀποδίδωσι καὶ ἀποτροπαῖς may actually have been an exegetes (most probably Kleidemos). The most important fact we learn is that Ph. (as would be expected considering Περὶ μαντικῆς) treated the significance of the days systematically: Schol. *Opp.* 763 contains remains of such a systematic treatment.

Plutarch (*Kamillos* 19.6: ὡς ἡμῖν ἐν τῷ Περὶ ἡμερῶν) wrote a Περὶ ἡμερῶν which the context shows was concerned with general historical events under their separate days. But when an author cites his source several times and specifically names the work, strong evidence is required before we impeach his authority. Laqueur, *RE* s.v. Philochoros (1938) 2435–2436, treats the nature of the Περὶ ἡμερῶν, showing that Philochoros treated days systematically. Whether Proklos used the Περὶ ἡμερῶν of Philo-

shows that the opposite was true of Athens. This again seems to me certain."

choros or of Plutarch, both are good sources, much better than the σχολικὰ ὑπομνήματα of the *Clouds*, which, I believe, omitted a line. It stands to reason that the Head of the Academy, one who "presided over the School for forty-seven years" (J. E. Sandys, *A History of Classical Scholarsh* I [1906] p. 372), had access to libraries in Athens and must have used the work he quotes.

A. Proklos on Hesiod *Op.* 765–768. Proklos, who wrote an *Outline of Astronomical Theories* and a commentary on Ptolemy's *Tetrabiblos*, says the true synod was sometimes on the 30th day, sometimes on the 29th when the day before is omitted by the Athenians.[11] The passage is a commentary on Hesiod's statement at the beginning of the "Days" that "the thirtieth day of the month is best for one to look over the work and to deal out supplies" (Loeb). Proklos is clearly referring to a lunar calendar, not a schematic one.

Meritt's argument, repeated several times but at greatest length in *Hesperia* 33 (1964) 3–4, is that the words ὑπὸ 'Αθηναίων are a gloss to be deleted. He uses the subjective argument of non-suitability to subject matter, which would eliminate hundreds of scholia and even modern studies,[12] writing, "The omission of Hesiod's

[11]Proklos was born in Byzantion and his horoscope fixes the date as February 8 412. He died in Athens in 485. O. Neugebauer, *A History of Ancient Mathematical Astronomy* 2 (1975) 1031–1036, gives a summary of his work. He wrote, "Proclus, in spite of all his philosophical prejudices had sound astronomical knowledge as is shown by his 'Hypotyposis'. This work provided the reader with a good introduction to Ptolemaic astronomy—one might say, the first and last summary of the contents of the Almagest from antiquity."

[12]M. L. West, *Hesiod Works and Days* (Oxford 1978) reminds us that there was a fuller text of the scholia by Proklos (p. 245), who wrote a continuous commentary on the poem. His exegesis may be incompatible with a text (p. 283), but that has nothing to do with its veracity. Meritt devotes space to the irrelevant fact that Pertusi misaccented ὅτε which has nothing

29th day was not brought about by anything done in Athens, and the absurdity of saying so condemns the tag ὑπὸ 'Αθηναίων." Proklos lived much of his life in Athens as Head of the Academy and might be expected to refer to Athens at a time when some Greeks may have been using the Julian calendar. He and other scholiasts on Hesiod's lines are writing about how people interpret ἡ σύνοδος. Meritt disregards the context, similar comments by other scholiasts, and misinterprets the complexity of Hesiod's calendar without any reference to the large bibliography.

On our passage, Walsh (*ZPE* 41 [1981] 119) writes, "It is true that the text can include the phrase ὑπὸ 'Αθηναίων, but, as will be seen, this evidence is nothing more than an unreliable conflation of the two calendric systems current in Greece." He writes "can include" because Meritt had tried to impeach the passage by noting that two MSS. (ZB) omitted ὑπὸ 'Αθηναίων. Pertusi determined the text on the basis of seven manuscripts. There are omissions in various places in all the MSS.[13] Walsh's two calendric systems are nonsense (pp. 109–110), that one used backwards count and ἕνη καὶ νέα for the last day, the other forward count and τριακάς. Meritt's game of terminology is corrupting.

B. The second scholion of Proklos is more troubling because it is a commentary on probably the most discussed word in the *Op.*, Hesiod's τρισεινάδα in line 814. Line 814 of the *Op.* reads as follows:

Παῦροι δ' αὖτε ἴσασι τρισεινάδα μηνὸς ἀρίστην

---

to do with what Proklos wrote. He also notes that a δέ is omitted in two manuscripts.

[13]See the stemma in Pertusi, *Scholia Vetera in Hesiodi Opera et Dies* (= *Pubblicazioni dell' Università Cattolica del S. Cuore* 53 [1955]) p. xx. B is derivative of Z, both much later than A from which they are derived: so the omission is insignificant. A is of the tenth century. B (XV) and Z (XIV) are much later. For the seven MSS., see Pertusi, p. 206.

Hesiod's calendar is discussed in M. L. West in his 1978 commentary on the *Op.* on his pages 349–350, Proklos' scholia on his pages 68–69. A. E. Samuel, in his article, "The Days of 'Hesiod's' Month," *TAPA* 97 (1966) 427, wrote, "Τρισεινάδα, which may mean 22, 27, or 29." T. A. Sinclair, in his 1932 edition of the *Op.* p. 87, cited five authorities as favoring 29, and five, including himself, as preferring 27. In his Müller's *Handbuch* (I.7) volume (1972) 66–68, Samuel showed that there were at least three different methods of counting the days of the month, adding that the term τρισεινάδα in line 814 fits none of the three. Paul Roesch, "Le calendrier d' Hésiode," *Stele. Tomos eis Mnemon Nikolaou Kontoleontos* (Athens 1980) 26–32, charts three plus systems used by Hesiod, taking τρισεινάς to be the 29th, "le 9 de la troisième décade." Since the word occurs in no calendar, he questions whether it was not "une appellation locale." It is to be noted that the text refers to the same day in different systems.[14] F. Solmsen, *TAPA* 94 (1963) 293–320, argued that the calendric part is the work of a number of poets, and this thesis was accepted by Samuel (p. 66 in the *Handbuch*). The only month named in the work (*Erga* 504 ff.) is not attested for Boiotia. West says there are sporadic examples of a forward count, such as Plutarch says was the case at Athens in a pre-Solon calendar.

To turn to the text of Proklos on τρισεινάδα, any interpretation that has Proklos say in one passage (765–768) the Athenians

[14]A. Lardinois, "How the Days fit the Works," *AJP* 119 (1998) 319–336, provides four pages of bibliography on the *Op.* He concludes (p. 324), "This 'multiculturalism' of the Hesiodic calendar can perhaps best be explained as another indication that Hesiod in the Days, as elsewhere in his poetry, is trying to establish a Panhellenic point of view, which transcends any one local tradition (Nagy [1982] 1990 and 1989)," a statement I find hard to accept.

omitted Day 29, and in another Day 22 or 27, is suspect. What I believe is the correct interpretation relieves Proklos of any contradiction and has Day 29 as the omitted one. F. A. Paley in his edition of Hesiod (London 1861) has the correct meaning (p. 102): "Proclus explains it thus, 'Few call the real twenty-ninth (so he explains τρισεινάδα, v. 814) by the name of twenty-ninth,' viz. because some, as the Athenians, counted the days of the last decad backwards, and so made the 'ninth' of that decad what was really the second." Paley is certainly correct, and Proklos continues τὴν πρὸ τριακάδος, the meaning of which is not in doubt. Numerical ligatures are found at times in all the MSS. Thus, in the following scholium (no. 817a), we read in Pertusi's apparatus (p. 224): "εἰκοστὴν καὶ ἐνάτην scripsi: κθ΄ codd." Alphabetical numerals were common in scholia and lexicographers, as well as in the great calendar list from Rhodes (*IG* XII.1.4), where the notation for days in the last decade is the following: κα (21), κθ (22), κη (23), κζ (24), etc. For the scholia of Hesiod this is shown graphically in the texts published by H. Schultz, "Die handschriftliche Überlieferung der Hesiod-Scholien": *Abh. Göttingen* 12.4 (1910). See in particular Schultz's index s.v. Zahlzeichen (p. 101), where numerical notations are used for days α to λ. Proclus was clearly referring to the day κβ΄ (29), or, as the Athenians called it in inscriptions with backward count, δευτέρα φθίνοντος. In short, Proklos wrote κβ΄ (29th day with backward count), which in the archetype of our MSS. was transcribed as δευτέραν εἰκοστήν.

In summary of this section on Proklos. The texts are sound, the Athenians omitted Day 29. I have presented the case for Proklos having an excellent source for the historical Athenian calendar, but an alternative answer, I suggest, is that he was referring in some passages to the calendars of his day. Simone Follet, *Athènes au II*[e] *et au III*[e] *siècle* (Paris 1976), in the course of her study of the Athenian calendar, concludes (p. 358), "Le calendrier lunaire a subsisté au moins jusqu'au IVe siècle."[15] When Proklos uses the present tense, he is not referring to pre-Solonian or pre-Metonic calendars, as Meritt and Walsh have it, but to the calendar of his day. As Head of the Academy in the mid-400's, he was writing for a general audience.[16]

5. Moschopoulos on Hesiod *Op.* 763 (= Gaisford, second ed. p. 417): – – – οὐχ ὑπεξαίρεσιν, ὡς 'Αθηναῖοι δηλονότι, ποιοῦντες· οὗτοι γὰρ τὴν πρὸ τριακάδος ὑπεξαιροῦντες ἐνάτην καὶ εἰκοστὴν τὴν τριακάδα καλοῦσιν.[17] The Athenians subtract the day before the 30th and call the 29th the 30th.

Moschopoulos possessed the writings of Proklos, author of several works on astronomy; so when we find the phrase ὡς 'Αθηναῖοι we can be sure that the same attribution was in Proklos, and attempts to get rid of the phrase in the latter are aborted. Walsh simply makes the pronouncement that the whole "scholiastic tradition" is to be condemned. He writes in part:

> In the scholion to verse 765 the word ἐξαίρεσις "extraction" was used to describe the process of omitting a date. In Moschopoulos' note the process is called ὑπεξαίρεσις 'subtraction'. The confusion in terminology reveals much about the failure, in the scholi-

[15]The scholiast on Aratos *Ph.* 740 writes, μῆνα δὲ λέγομεν τὸν ἀπὶ συνόδου σελήνης καὶ ἡλίου χρόνον ἐπὶ σύνοδον ... τούτῳ δὲ τῷ μηνὶ ἐχρῶντο πρὸς τὴν τῶν πολιτικῶν ἡμερῶν διαγωγήν· καὶ νῦν ἔτι χρῶνται πολλοὶ τῶν Ἑλλήνων, but I do not know the date of the scholion, which has many Greeks using a lunar calendar, presumably at some late date. Cf. Mommsen, *Chronologie* (1883) 62–63.

[16]Walsh's argument is that Proklos confused a forward counting calendar, which was pre-Solonian, with a backwards counting one. Moreover, he attributes Proklos' alleged error to the Solonian, or Meritt's pre-Metonic calendar, as if Proklos more than a thousand years later possessed such a copy, and, furthermore, ignored subsequent Athenian count. The whole thesis is unsubstantiated. The combination of Proklos and Moschopoulos leaves no doubt but that the Athenians omitted Day 29.

[17]Pertusi did not include scholia from Tzetzes and Moschopoulos.

> astic tradition, to make the necessary distinctions between these two different calendric systems. In the Athenian backward counting calendar the process of omission should properly be called ἐξαίρεσις because a date was extracted from the heart of the hollow month, ...

He ignores the fact that both Proklos and Moschopoulos were in agreement in using the ὑπο– compound for the Athenians' omission of a day, i.e. a day at the end of the month. Proklos wrote, ὑπεξαιρεῖται ἡ πρὸ αὐτῆς ὑπὸ 'Αθηναίων. M. L. West (p. 70 of his edition of Hesiod *Op.*) writes, "There is less to remark about the commentary of Manuel Moschopoulos, written sometime between about 1290 and 1310. Moschopoulos was far better equipped to understand Hesiod than either Proclus or Tzetzes; he seems a model of common sense in comparison."

6. Geminos. An early passage in Book 8 of the *Eisagoge* is the only source cited by Meritt for his theory of alternating full and hollow months. The passage in question will be discussed in Chapter V. Here we are concerned only with an "omitted day" passage (8.55–56) to which reference was made in Chapter II pp. 27–28. Heath's translation of the passage is:

> And they included 110 hollow months in the 235 months for the following reason. As there are 235 months in the 19 years, they began by assuming each of the months to have 30 days; this gives 7,050 days. Thus, when all the months are taken at 30 days, the 7,050 days are in excess of the 6,940 days; the difference is (110 days), and accordingly they make 110 months hollow in order to complete, in the 235 months, the 6,940 days of the 19-years period. But, in order that the days to be eliminated might be distributed as evenly as possible, they divided the 6,940 days by 110; this gives 63 days. It is necessary therefore to eliminate the [one] day after intervals of 63 days in this cycle. Thus it is not always

> the 30th day of the month which is eliminated, but it is the day falling after each interval of 63 days which is called ἐξαιρέσιμος.

Meritt's chart of the exairesimoi days in this 19-year cycle attributed by Geminos to the astrologoi Euktemon, Philippos, and Kallippos is published in *AE* 1968 p. 98. His theory was that κατὰ θεόν = κατὰ Μέτωνα, and that this calendar of Geminos 8.55–56, with full months omitting every 64th day, served as a regulatory calendar for the κατ' ἄρχοντα one according to which all festivals were performed; see the table of triple dates p. 2–3. Hence if there was no tampering, the 64th day was omitted. But in his treatment of calendar equations in individual years, Meritt regularly posits that there were hollow months with the omission of Day 21/22. Why the Athenians fashioned two such contradictory calendars in which the days of the year rarely coincided and how calendric authorities administered the two is never disclosed. The effect and apparent purpose of the creation of this strange and impossible theory is to maintain that Skirophorion was sacrosanct and that we do not have evidence that tampering extended into this month. I observe that when Meritt applies the 64-day calendar, it is in the context of complete months; see, for example, *AE* 1968 p. 99. When he applies the Day 21/22 calendar, the context is days. Whatever the explanation, the results of the advocacy of this dual calendric theory is reflected in Meritt's 1977 archon table (*Historia* 26 161–191).

Conclusions to Chapter III

No rationale has been offered for how grammarians and scholiasts would have known of a calendar in use only from Solon to Meton and then reported it as the Athenian calendar of all ages.

Just as Scribe B of Pollux jumped a line in reading alphabetical numerals, so with the source of the scholion in Aristophanes *Clouds* 1131.

Proklos, Head of the Academy in Athens for many years, expressly says δευτέρα φθίνοντος was the omitted day in Athens. He possessed Philochoros' Περὶ ἡμερῶν and may have reported on a calendar which persisted to his day.

Moschopoulos confirms that what Proklos wrote was that the Athenians omitted δευτέρα φθίνοντος.

The theory that the Athenians adopted a regulatory calendar of full months with the omission of Day 64, designating it as κατὰ θεόν in contrast with a festival calendar (κατ' ἄρχοντα) omitting Day 22, is a calendric monstrosity, apparently devised in order to avoid the admission that the month Skirophorion (XII) was subject to tampering.

There is no literary evidence that Day 22 was omitted, and the advocates of such have never adduced any.

IV

LUNAR VS. LEAP-DAY CALENDAR

A proper title for this chapter is difficult. In *Calendars of Athens* (1947), we advocated a lunar calendar. Subsequently, Meritt responded by claiming in all of his many publications that the Athenians used a calendar of alternating full and hollow months. The alternation of 6 full with 6 hollow months gives a moon-year of 354 days; but the true moon-year exceeds this by 0.36707, T. L. Heath, *Aristarchus of Samos* p. 288, says, and, hence, after about 2 3/4 moon years, a day would have to be added in order to keep the months in harmony with the phases. The calendric difference after **n** years would be **n** × 0.36707. After nineteen years, the difference would be 6.97433. No scholar has ever charted such a calendar for the Athenians, although several have given lip-service to the phrase. Meritt's theory of leap-days for his calendar of alternating months will be discussed later in the chapter. First, we turn to our theory of a lunar calendar.

LUNAR CALENDAR

1. The phrase κατὰ θεόν. My collaboration with Otto Neugebauer after the Second World War sprang from a conversation in which he explained that the term κατὰ θεόν, found in the inscriptions listed in Table 1 Chapter 1, was equivalent to "According to the moon" (κατὰ σελήνην) and denoted a lunar calendar. We wrote (*The Calendars of Athens* [1947] 16), "The name κατὰ θεόν can be translated 'according to the moon.' This is shown, e.g., by the expression κατὰ θεὸν νουμηνία in *P. Mag. Par.* I and similar expressions in the astronomical literature, e.g., Cleomenes where sun and moon are frequently called θεοί." As noted in *ZPE* 128 (1999) 84, E. G. Turner and O. Neugebauer, commenting on the

phrase κατὰ σελήνην in a papyrus (*Bulletin of the John Rylands Library* 32, No. 1 [1949] 9), write, "The addition of *kata selenen* obviously means that we are dealing here with phenomena of the real lunar calendar." In Josephus (*AJ* 2.318; 3.248) dates were given by lunar reckoning (κατὰ σελήνην). Pouilloux demonstrated that kata theon and kata selenen were synonymous in Philo: *REA* 66 (1964) 211–213. J. and L. Robert (*REG* 78 [1965] p. 94 no. 88), in commenting on Pouilloux' article, added *IG* VII.4135 line 11, where the phrase kata theon is used in referring to the same date in the calendars of Boiotia and Delphi. This Boiotian inscription is studied by Paul Roesch, *Études Béotiennes* (Paris 1982) 60, who presented four examples of kata theon in the Boiotian calendar, translating the term as "le calendrier lunaire."

That Selene was early worshipped as a goddess appears from Homeric Hymn 32, titled Εἰς Σελήνην, where she is also called Μήνη; cf. Pindar *Ol.* 3.20. She had a daughter Pandia (line 15). Pandia is the name of an Athenian festival, which some think, apparently incorrectly, came at the end of the Great Dionysia; see the commentary of Allen, Haliday, and Sikes, *The Homeric Hymns*[2] (1963) p. 435. See also L. Deubner, *Att. Feste* 176–177. Macrobius, in his *Commentary on Scipio's Dream* 2.11.6, says, "In Greek the word for the moon is mene." Μείς, μηνός is used hundreds of times for the month. In 2.28, Thucydides adds κατὰ σελήνην to νουμηνία in referring to the solar eclipse of 3 August 431. Dover (*HCT* IV.265) writes, "He thereby indicates his awareness of the possibility that moon and month could on occasion be out of step." If days had been suppressed or intercalated, the deviation was from Selene, not from a cyclic, or "Metonic", calendar of alternating full and hollow months, or from one omitting every 64th day. The chorus leader in Aristophanes *Clouds* 614 ff. attributes to Selene the statement (tr. Henderson): "She says that though she does you other favors too, you don't keep track of your dates correctly, but scramble them topsy-turvy, so that the gods scold her, she says, every time they're misled about a dinner and go home having missed the festival that was specified in the calendar", concluding with the line κατὰ σελήνην ὡς ἄγειν χρὴ τοῦ βίου τὰς ἡμέρας. The thrust of the passage is that the untampered calendar was "according to the moon."

One of our most informative literary passages about the Athenian calendar is found in Plutarch *Solon* 25.3 (Loeb tr.):

> Observing the irregularity of the month, and that the motion of the moon does not always coincide with the rising and setting of the sun, but that often she overtakes and passes the sun on the same day, he ordered that day to be called the Old and New, assigning the portion of it which preceded the conjunction to the expiring month, and the remaining portion to the month that was just beginning. He was thus the first, as it would seem, to understand Homer's verse, which speaks of a day when
>
> "This month is waning, and the next is setting in,"
>
> and the day following this he called the first of the month. After the twentieth he did not count the days by adding them to twenty, but by subtracting them from thirty, on a descending scale, like the waning of the moon.

One may question Plutarch's sources for any knowledge of Solon's calendar, but he lived in Greece and knew the calendar of his day, the author of a Περὶ ἡμερῶν. His phrase ὥσπερ τὰ φῶτα τῆς σελήνης, mistranslated in the Loeb, is more explicit than κατὰ σελήνην and leaves no doubt that the reference is to the cycle of the moon in a true lunar calendar.

A clinching argument that κατὰ θεόν = κατὰ σελήνην occurs in one of our best sources, Aristotle *Ath.Pol.* 43.2, who writes of the Athenian calendar: κατὰ σελήνην ἄγουσιν τὸν ἐνιαυτόν. In a third-century decree from Stymphalos (*IG* V2.357 line 18), we have δεκάται κατὰ σελ[άναν]. The two kata-phrases are identical (moon dates). That κατὰ θεόν and κατὰ Σελήνην are synonymous may be inferred from an arbitration treaty between Messenia and Lakedaimonia ca. 140 B.C., where a common date is given κατὰ σελήνην (*SIG*³ 683.44–46): καὶ ἡμέραι ἐνδεκάτηι κατὰ σελήνην ἀφ' ἧ[ς ἡμέρας τὸ] δόγμα ἐγένετο. As noted in *ZPE* 128 (1999) 83, we have in *IG* II² 951 add. the record of an oath taken by emissaries of three far-removed states, Athens, Ambrakia in the north, and Akarnania in the west about 166 B.C., which gives the following opening lines:

> ἐπὶ Νικοσθένου ἄρχον[τος μηνὸς Θαργηλιῶνος]
> πέμπτει ἀπιόντος ὡ[ς 'Αθηναῖοι, ὡς δὲ 'Αμ]–
> βρακιῶται ἐπὶ γραμ[ματέως . . . .⁹. . . . . μ]ηνὸς [Φοι]–
> νικαίου πέμπτει ἀ[πιόντος, ὡς δὲ 'Ακα]ρνᾶνε[ς]
> ἐπὶ στρατηγοῦ Χρεμᾶ [μηνὸς Ἑσπερι]νοῦ? τετρά[δι]
> ἀπιόντος, ὅρκος . . .

Here the day at Athens and Ambrakia is given as XI 26, at Akarnania in the west as XI 27. This uniformity suggests that these three states were using a lunar calendar, and if these three probably all of Greece.

Other references are to the Greeks in general and do not exclude the Athenians. For example, Aratos, who studied at Athens and flourished in the middle of the third century B.C., is the author of a poem *Phaenomena*. The first and longest part is the versification of a poem treatise by Eudoxos describing the constellations. The second part (733 ff.) deals with weather signs and is perhaps derived from Theophrastos, in which he enumerates certain days of the month with weather prognostications; see G. J. Toomer, *OCD*³ (1996) 136–137. In lines 733–738, he wrote (tr. Douglas Kidd [1997]):

> Don't you see? When the moon with slender horns is sighted in the west, she declares a waxing [735] month; when the first light shed from her is enough to cast a shadow, she says she is entering on the fourth day; eight days she indicates at half-moon,

mid-month when she is with full face. As she continually changes her aspect with different phases, she tells which day of the month is taking its course.

The reference to the month being determined by the phases of the moon is clearly applicable to the Greek calendars in general. Cf. Mommsen, *Chronologie* p. 58 n. 2 and p. 183 n. 2. Indeed, the scholiast on line 735 writes, ἐμέτρουν γὰρ οἱ Ἕλληνες τοῦ μῆνας πρὸς τὸν τῆς σελήνης δρόμον, Αἰγύπτιοι δὲ πρὸς τὸν τοῦ ἡλίου. These scholia on Aratos, which derive from Theon of the fourth century, are said to be generally of a high order. Aratos is a good witness for his own period, when a new religious movement involving the worship of the sun, moon, and stars had spread over the ancient world.[1]

Geminos, after treating the Egyptian calendar in 8.16–24, introduces the various cyclic calendars of the astronomoi with the statement in 8.25: οἱ δε Ἕλληνες τὴν ἐνιαντίαν γνώμην ἔχοντες τοὺς μὲν ἐνιαυτοὺς καθ' ἥλιον ἄγουσι, τοὺς δὲ μῆνας καὶ τὰς ἡμέρας κατὰ σελήνην. Again, the Athenians are not excluded. Other passages in Geminos, which Meritt makes the foundation for his calendar, will be studied below.

R. Meiggs, *CR* 77 (1963) 334, in reviewing Meritt's *Year*, spontaneously uses the phrase "dates κατὰ θεόν (moon dates)." Although the phrase was one of the cornerstones for our theory of a true lunar calendar in *Calendars of Athens* and in spite of hundreds of pages devoted by Meritt and now by Woodhead in argument against our position, never does either scholar duplicate what Meiggs recognized was obvious, that κατὰ θεόν = κατὰ σελήνην, "moon dates." Nor do they refer to the literary passages which have been cited here as supporting this position. All is ignored and goes by default. We are told that the κατὰ θεόν dates

[1]See A. M. J. Festugière, *La révélation d'Hermès trismégiste*, Vol. II: *Le Dieu cosmique* (Paris 1951).

are those of the zodiacal calendar attributed to Euktemon and others (and Meton?) by Geminos in 8.55–56, a calendar of full months with the omission of every Day 64, but Meton was not a θεός. At other times, we are told that it is an alleged calendar of Geminos 8.3 (μῆνες ἐναλλὰξ πλήρεις καὶ κοῖλοι), to be discussed below. For Woodhead, the phrase means κατὰ Meritt. The view that the θεός calendar gives "moon dates" and that the theos is Selene, has not been addressed, much less refuted. One never finds the word Selene in all of Meritt's writings. One of the strangest arguments that the words kata theon could not refer to an observation calendar is advanced by Merritt in *TAPA* 95 (1964) 234. In the Eretrian inscription, *IG* XII.9.207 lines 59–61, we are told that the Dionysiac performers who wish to sign contracts are to present themselves in Chalkis before the twentieth day kata theon (π[ρὸ εἰκάδων τ]ῶν κατὰ θεόν). Meritt writes in part:

> It would have been virtually impossible for the artists who wishes to present themselves in Chalcis to know the 20th day κατὰ θεόν if all their several calendars κατὰ θεόν were based on observation of the new moon in their several cities, and neither the Histiaean from Oreos nor the Carystian from Carystus could have known when the Chalcideans "observed" the new lunar crescent in Chalcis.

A calendar κατὰ θεόν is epigraphically attested for Athens, Boiotia and Delphi as well as for Euboia, and one κατὰ σελήνην for Athens, Stymphalos (Arkadia), Messenia, and Lakonia. Since in the article in question, Meritt believes that the κατὰ θεόν calendar was that of Meton, as attested for three astrologoi in Geminos 8.55–56 with all months full and every Day 64 omitted, he would have it that in effect virtually all of Greece was using a calendar omitting every 64th day, a calendar which Neugebauer, Toomer, Aujac, and others find impractical even for astrologoi. Meritt writes (p. 242), "Meton's calendar we know was laid out on a plan by which all months were first assumed to be full, of thirty days, and then one day was omitted after every 63. This had the effect of giving alternating months with an occasional double full." Thus, for Meritt, a calendar of full months omitting every 64th day regardless of the count becomes one of alternating full and hollow months omitting Day 21/22. Confusion reigns, and that for all Greek calendars, all arising, I suggest, from a misunderstanding of observation.

The length of the island of Euboia is about ninety miles. The implication is that observers in the four cities would have difficulties in agreeing on the date while looking at the same moon. When the moon is closest to the earth (perigee), it is 221,463 miles away. The difference in the angle of viewers separated by 90 miles or less is of the slightest. Allowing for different horizons, the difference between Chalkis and Eretria, 22 kilometers apart, could not be as much as a day, and so with the other cities. Since Samuel (*GRC* p. 98) says the cities did not share a common calendar, the date was given kata theon. This is clear evidence for a lunar reference.

Many peoples, surely numbering in the millions, have lived with lunar calendars. The Mosaic law bound the beginning of the new month to the new crescent. Alec Waugh, *National Review* January 26, 1971, has an interesting note on his experience about the observance of the crescent for the first of Shawwal after the last day of Ramadan in Tangier. Richard A. Parker, *The Calendars of Ancient Egypt* (Chicago 1950), at the beginning of his Chapter 1 (p. 9) wrote:

> All that we know of ancient and modern time-reckoning leads to the conclusion that lunar months begin with some *observable* phase of the moon. "As always," says Nilsson, "the concrete phenomenon is the starting point." Most peoples (both ancient and modern) who use a lunar calendar start the month with the new crescent; a few count from full moon; while two East African tribes, the Masais and the Wadschaggas, begin with the moon's invisibility. All these starting points, as well as one other, con-

> junction, have been maintained for the Egyptian lunar month by various scholars from as recent a year as 1932. It is the purpose of this chapter to demonstrate that the ancient Egyptians, like the Masais and the Wadschaggas, began their lunar month on the morning of the day when the old crescent of the moon was no longer visible in the eastern sky before sunrise.[2]

E. J. Bickerman, *Chronology of the Ancient World* (1968) 17, writes in more general fashion:

> Almost every people of the earth have used the lunar phases for measuring time: *luna regit menses* (Ovid, *Fasti* III, 833). The Greek word μήν (*men*) and our term 'month' equally point to the moon. The same is true for the terminology of the Semitic languages. For instance, in Hebrew the word *yerah* means 'moon' and 'month', and the other word for 'month', *hodesh*, properly signifies the 'new' crescent.
>
> As a matter of fact, almost all the peoples of the Mediterranean world, the Celts (Plin. *N.H.* xvi, 44), the Germans (Tac. *Germ.* 11), as well as the Hebrews and the Babylonians, began the month at the apparition of the young crescent, as the Islamic peoples still do today for their religious calendar: the new moon signals the longed-for end of the fast month (Ramadan). The beginning of the month was sometimes publicly announced (for Greece

[2]Since the epoch of the day was sunrise in Greece as in Egypt, and Day 29 was omitted, I have urged in several publications, most recently in *ZPE* 49 (1992) 263–264; 128 (1999) 87, that the Athenians determined a hollow month by the invisibility of the old crescent in the eastern sky. Needless to say, any calendar depended originally on observation. Leo Depuydt, *Civil Calendar and Lunar Calendar in Ancient Egypt* (1997) 27–28, explains that when the epoch of the day was morning, peoples observed the moon in the morning to begin the lunar month. Add to my testimonia Geminos 6.1: Ἡμέρα λέγεται διχῶς, καθ' ἕνα μὲν τρόπον χρόνος ὁ ἀπ' ἀνατολῆς ἡλίου μέχρι δύσεως, καθ' ἕτερον δὲ τρόπον ἡμέρα λέγεται χρόνος ὁ ἀφ' ἡλίου ἀνατολῆς μέχρις ἡλίου ἀνατολῆς. Cf. James Evans, *The History and Practice of Ancient Astronomy* (1998) 239.

cf. Nilsson, *Kalendar* 29). In early Rome, the *pontifex minor* observed the sky and announced the new moon.

Atmospheric conditions may obscure the moon at the end of the month, but experience would dictate that the month should begin anyway. A. E. Samuel, *GRC* 15, writes, "Poor visibility may cause a calendar month to be longer or shorter than an observation month, by a maximum of one day, but visibility conditions can only delay the start of a calendar month, causing it to be behind an observation month, and never can accelerate the beginning of a calendar month ahead of observation." At the turn of the last century, Frazer could see Aigina from Athens. I doubt that in antiquity in June and July there were many years in which the moon was not visible for observation at the end of Skirophorion for Hekatombaion 1. The same would be true of the four Euboian cities. The winter months might present problems.

One of our earliest inscriptions (*IG* I^3 6.B) prescribed periods of days for sacred truces using lunar dates which must have been equivalent to kata theon ones and known to all Greeks. Such early terms as δεκάτη ἱσταμένου and διχομηνία (c. 460 B.C.) surely suggest a lunar calendar. Days so designated would be meaningless unless on average they fell on the designated ones, particularly since they were directed at a Greek-side audience. The virtue of the kata theon, or kata Selene, calendar was its universality, used in designating future dates or recurring events.

The four cities of Euboia certainly did not share Meritt's alleged "Metonic" schematic calendar. Rather, about a century earlier than attested at Athens, the Euboians were distinguishing a true lunar calendar from local festival calendars subject to embolimoi/exairesimoi days, which the Athenians called κατ' ἄρχοντα.[3]

[3]As Geminos tells us, the Greek calendars regulated festivals, sacrifices, games, and such activities, which all city-states observed. The Athenians alone used the phrase κατ' ἄρχοντα to designate irregularities, but such

We learn something about the calendar from passages designating places of observation with Jacoby's commentary. Strabo (2.5.14.119) says Eudoxos, who flourished 368–365, had an observatory (σκοπή) at Knidos not much higher than the dwelling-houses where he was able to sight the star Kanopos on the southern horizon. He also had observatories (σκοπαί) at Heliopolis in Egypt where he could observe movements of the heavenly bodies: Strabo 17.1.30.807. Plutarch (*Dion* 29.2) says that Dion in 357 delivered a speech from Achradina at Syracuse on a heliotropion which Dionysios had set up. Moschion, quoted by Athenaios 5.207e, says the ceiling of Hieron's ship (by Archimedes) was a concave dial made in imitation of the heliotropion on Achradina. Theophrastos (*Sign.* 1.4[4]) names three places used for observations, including Mount Lykabettos at Athens by the astronomos Phaeinos, who, he says, was a teacher of Meton. He adds, καὶ ἄλλοι δὲ τὸν τρόπον τοῦτον ἠστρολόγησαν. As early as 428/7, Aristophanes in his *Daitaleis*, in a corrupt fragment (not in Kock or Edmonds) from the Vaticanus of Achilles' *Isagoge* 29, referred to a sun-dial: ὁ δὲ Ἀριστοφάνης ἐν τοῖς Δαιταλεῦσιν ἐπὶ ἡλιοτροπίου (Maass, ἢ ἐπεὶ ἡλίου V) τέθεικε τὸ ὄνομα τοῦ πόλου καὶ θηλυκῶς· «πόλος τοῦτ' ἔστιν +ἡλιοκολωνῶ ἐν ὧ / σκέπτουσι (σκοποῦσι Wil) τὰ μετέωρα ταυτὶ καὶ τὰ πλάγια ταυτί». Again, in the *Gerytades* (409/8), he refers to the πόλος

---

must have occurred in all states, which explains many calendric differences between states; see, for example, *Ancient Athenian Calendars on Stone* (1963) 326–328. The Athenian κατ' ἄρχοντα and κατὰ θεόν dates illustrate irregularities common to all states, although not so designated. As at Athens, single dates are κατ' ἄρχοντα. Of course, they may also be κατὰ θεόν.

[4]According to Neugebauer, *HAMA* 2.739, Rehm suggested Euktemon as the author of this work. Regenbogen, *RE* Suppl. 7 (1940) 1412–1415, collects several possibilities.

(frg. 163 K).[5] What Meton dedicated at Athens is disputed. The Ravennas scholiast to Aristophanes *Aves* 997 says simply of Meton: φησὶ δὲ Καλλίστρατος ἐν Κολωνῷ ἀνάθημά τι εἶναι αὐτοῦ ἀστρολογικόν. The Venetus scholion is much longer and is published by Jacoby, *FGrH* 328 frg. 122. It cites Philochoros as saying Meton's ἡλιοτρόπιον was on the Pnyx. Both A. H. Sommerstein (pp. 264–265) and N. Dunbar (pp. 554–555) in their recent editions of the *Aves* comment on the scholia. Dunbar believes Meton set up two sundials. Jacoby's comments on the Philochoros fragment (122) cover several pages (3 B Suppl. 1 pp. 496–498, Suppl. 2 pp. 401–403). He argues that Meton's meteorological calendar of the Enneakaidekaeteris was not an anathema,

[5]The terms polos and gnomon are discussed in *CR* 54 (1940) by E. Powell (p. 69) and at greater length by D. S. Robertson (pp. 180–182). For bibliography on sundials, including the book of Sharon Gibbs, see *ZPE* 128 (1999) 91–92. James Evans, *The History and Practice of Ancient Astronomy* (1998) 129 ff., provides sketches of a variety of ancient sundials. For Greek "shadow tables" to determine the hour of the day, see Neugebauer, *HAMA* 2.736–740; 208 ff. Cf. James Evans, *The History and Practice of Ancient Astronomy* (1998) 61–62; and Hultsch, *RE* s.v. Gnomon (1912) 1500–1502, citing Herodotos 2.109 (Loeb tr.): "The Greeks learnt the art of measuring land; the sunclock (πόλον) and the sundial (γνώμονα) and the twelve divisions of the day, came to Hellas not from Egypt but from Babylonia." Demokritos wrote a treatise on the κλεψύδρα as well as on the παράπηγμα: Diogenes L. 9.48. *Isaiah* 38.8 says, "I will make the shadow set by the declining sun on the dial of Ahaz turn back ten steps. So the sun turned back on the dial the ten steps by which it had declined." Aristophanes *Eccl.* 652 is the earliest Athenian author to refer to the shadow on the dial. B. B. Rogers, in his commentary, says he experimented and claims the gnomon at Athens was one foot high. Probably, the best collection of testimonia remains that of Ardaillon in DS *Dictionnaire* 3.1 s.v. Horologium 256–264. Evans (pp. 95–98), citing Aratos, believes the Greeks told time at night by the risings of the zodiac constellations. He believes that from the fifth century on this was information the average person would have.

but a technical paper,[6] certainly not officially introduced when it appeared. This calendar of Meton is generally taken to be that of Geminos 8.50–56, attributed to Euktemon and two other astrologoi, in which all months were full with Day 64 being omitted. I support Jacoby's position, and add that a calendar omitting Day 64 was never made law by the Athenian ekklesia.[7] We have the

[6]Earlier, Kubitschek, *RE* s.v. Meton 1464.5–8, referred to it as "perhaps literary" citing Diodoros' reference (12.36.2) to it as γραφή. He says (1463.61) the omitted day should be after 63.36 days. Ginzel (*Handbuch* 2, 403–405) charted four different cycles, Ideler (intercalary = 3 5 8 11 13 16 19), Mommsen (3 6 9 11 14 17 19), Unger (3 6 8 11 14 17 19), and Schmidt (2 5 8 11 14 16 18). None of the four are in accord with that of Fotheringham, *Monthly Notices of the Royal Astronomical Society* 84 (1923/24) 383–387 (0 1 0 0 1 0 0 1 0 1 0 0 1 0 0 1 0 1 0), which was adopted by Meritt and Woodhead without any argumentation. For criticism of Fotheringham's calendar, see O. Neugebauer, *HAMA* 2.617. Meritt's Metonic cycle in *AE* 1968. 98 has his first exairesimos day as Metageitnion (II) 16 (30 + 15 is not 63) and a year of 383 days! All five theoretical reconstructions allow seven intercalary years; Meritt (*Year* 5) would allow eight. All five assume a fixed intercalary month; we know that the Athenian calendar had various intercalary months (see Table 3, p. 8). The comments of Van der Waerden (*JHS* 80 [1960] 176–179) against the Athenian adoption of this calendar are worth consulting. Meritt and Woodhead made no effort to respond to these and others. The task of any future scholar who theorizes that the Athenians had a fixed 19-year cycle is first to establish with some probability what the cycle was, which means clearing the slate of the series of asterisks for the character of years in the archon table published in *Historia* 26 (1977) 161–191, followed by Woodhead in *Agora* XVI. It will not do to say simply that modern scholars x and y (or a majority) claim such and such a year is ordinary or intercalary, always bearing in mind that what one archon or mason did, another might, and that single dates might be kat' archonta or kata theon. One may say positively, I think, that the Athenians never adopted the calendar of Geminos 8.50–56 with its cycles of 6940 days. Toomer, *Dictionary of Scientific Biography* IX (1974) 338, refers to it as "an artificial astronomical calendar."

[7]The parapegma of Euktemon or Meton, an inscription on stone for public use, was of a different order. It listed the risings and settings of a number of bright stars and star-groups for the purpose of predicting weather

phrase κατ' ἄρχοντα, not κατὰ ψηφίσματα or κατὰ Μέτωνα. It remained a theoretical approach for a series of 19-year cycles. I have referred above to the doubts of Neugebauer, Toomer, and Aujac about its practicality. No elective body could have been persuaded to legislate such a calendar for the regulation of its festivals, and Meritt's endorsement of such a calendar lacks all credibility,[8] and his claim that it amounted to a calendar of alternating

---

conditions and the like. It listed the days by months according to the passage of the sun through the zodiacal constellations, and is not to be confused with a calendar of 6940 days never inscribed on stone. See, for example, Pritchett and Van der Waerden, *BCH* 85 (1961) 32–47.

[8]See the table in *AE* 1968. 98. It comes as no surprise that David Lewis in *JHS* 83 (1963) 195 claims that the Aristophanic scholia are superior to Proklos, head of the Academy, as a source, and in *CR* 86 (1972) 259 that Meritt's interpretation of κατὰ θεόν = κατὰ Μέτωνα is correct. Despite the endorsement of Meritt and Lewis, other scholars have rejected the practicality of the "Metonic" calendar attributed by Geminos to Euktemon, Philippos, and Kallippos (8.50 ff.) for any civil calendar. Neugebauer (*HAMA* 2.617) wrote, "One can safely dismiss as completely unhistorical the concept of a calendar supposedly made for the astronomers alone but nowhere attested in astronomical records." As noted by K. von Fritz, *RE* s.v. 48, Phillippos of Opus (1974) 2357, Geminos (8.59) contradicts himself by attributing to Kallippos a different cycle. See also T. L. Heath, *Aristarchus of Samos* 295–296. If unhistorical for astronomers, a calendar which has 3 and 5 as consecutive days in one month and 19 and 21 in another month, etc. (*AE* 1968.98), was even more so for the festival calendar of any city-state, including Athens. Censorinus, in the *De Die natali* 18.8, gives quite a different account of the 19-year cycle (of Meton). After saying that it had 110 hollow and 125 full months, he continues, "so that hollow and full months did not always follow one another alternately, but sometimes there would be two full months in succession. For the natural course of the phenomena in regard to the moon admits of this, whereas there was no such thing in the octaeteris." See Dinsmoor, *Archons of Athens* 310. Because of the errors and faulty traditions in Geminos, documented by Manitius (pp. 252 ff.), I judge that Censorinus was right, although both are based entirely on secondary sources. But it cannot be repeated too often, these were not civil calendars, but ones made for astrologoi. Van der Waerden, following Fotheringham, tries to defend the

full and hollow months is utterly false. No orator, an ancient equivalent of a Meritt or a Woodhead, would have stood before an assembly and proposed a complete rescheduling of festivals with all receiving an omitted day somewhere in the course of the cycles. Meritt's theory was a base on which he pyramided his rejection of a true lunar calendar. We do not need to rewrite Mommsen and Deubner.

2. Years of 353 and 383 Days. Meritt, in accord with his theory that the Athenian calendar was one of alternating full and hollow months, argued throughout that the festival year was one divided into twelve or thirteen months which had 354, 355 (ordinary) or 384 (intercalary) days; see, for example, *Year* p. 6; *TAPA* 95 (1964) 200; *Eph. Arch.* 1968. 92. Any year of 353 (ordinary) or 383 (intercalary) days would disprove his theory and be in accord with a true lunar calendar.

In his *The Sacred and Civil Calendar of the Athenian Year* (Princeton 1975), Jon D. Mikalson devoted pp. 177–181 to a catalogue of decrees passed on Skirophorion (XII) ultimo, listing ten which preserved that date without restoration and eleven with restorations. He wrote (p. 181): "From the ten meetings attested for Skirophorion 30, more than twice as many as attested for any other day of the year, we may infer that there was a meeting of the Ekklesia on Skirophorion 30 every year. This is also indicated by the fact that no meeting is attested for Skirophorion 29."

Assuming that the prytanies of intercalary and ordinary years corresponded with the Aristotelian pattern given in *Calendars of Athens*, we can set up equations with the Skirophorion ultimo dates which establish the length of years. Most are agreed, I be-

---

Geminos passage as such (*JHS* 80 [1960] 170–171). One must not ignore the fact that other fifth-century cycles than those found in Geminos, and these with different lengths for the day, are attested (Philolaos, Oinopides, Demokritos): Neugebauer, *HAMA* 2.619–620.

lieve, that the length of prytanies was determined in advance at the beginning of the years and some regularity was observed; for intercalary years most are in agreement. In the period of ten phylai before 307/6 B.C., the last prytany was of 35 days in an ordinary year of 354 days if the festival calendar was of alternating full and hollow months, but if lunar of 35±1 (*Calendars of Athens* p. 36). In an intercalary year, the last prytany was of 38 days or of 38±1 (p. 37).

In the ordinary year which Kirchner dates about 330/329 B.C.,[9] *IG* II² 415 gives a firm equation without benefit of restoration:

Σκιροφοριῶνος ἕνηι καὶ νέαι, τετά[ρτ]ει καὶ τριακοστῆι

Skirophorion (XII) ultimo = Prytany X 34 = 353 day of the year

Through the date of our equation there were 7 29-day months and 5 30-day ones, which is common with a lunar calendar, but impossible in a rigid calendar of alternating full and hollow months.

Again, in the intercalary year 168/7 in the period of twelve phylai when the prytanies were of 32 days with the last 32±1, we have in *IG* II² 945 the following equation without benefit of restoration:

Σκιροφοριῶνος ἕνηι καὶ νέαι, μιᾶι καὶ τριακοστεῖ

Skirophorion (XII) ultimo = Prytany X 31 = 383rd day of the year

This count, too, indicates a lunar, not a schematic year. It would be rash to posit that with a predetermined prytany calendar all prytanies in intercalary years were not distributed equitably, 32 days each; cf. Meritt, *Historia* 26 (1977) 165.

The phrase Σκιροφοριῶνος ἕνη καὶ νέα προτέρα. As documented in Mikalson, *The Sacred and Civil Calendars*, pp. 178–

[9] M. B. Walbank, *Hesperia* 51 (1982) 44, suggests a "slightly earlier date", and identifies the hand with that of a new decree found in the Agora excavations; see *SEG* 32 no. 72.

179, the phrase occurs in *IG* II^2 486, *Hesperia* 7 (1938) 297, *IG* II^2 495, 496, 497, 597 add., and Meritt, *Year* 192–194. At the end of the year, the calendar was extended by one day and one day only. This is the only month for which the phrase is found without benefit of restoration. The significance, I suggest, is that observation after Day 28 in a calendar where the epoch of the day was sunrise determined the need for the addition of the extra day. Observation at the end of the year was made with particular care, and one expects to find a degree of uncertainty in an observation calendar. Such an addition is impossible in a calendar with strictly alternating full and hollow months. Because of the dates given in the prytany calendar in each equation, the years would have been of 353 and 383 days if the extra day had not been added, which is proof that a calendar of strictly alternating months was not used. Meritt and others have argued that this was impossible.

One other Day 30^1 has been restored in *IG* II^2 354, this time for the month of Elaphebolion (IX). The name of the month was not inscribed and the prytany date must be restored, which disqualifies the text from providing certainty. The calendar equation was read as ἕνηι καὶ νέαι π[ροτέραι, εἰκοστ]ῆι τῆς πρυτανείας. A photograph of a latex squeeze of the right margin showing the π, which should be dotted, was published in *Ancient Athenian Calendars on Stone* pl. 22a. In *Calendars of Athens* 51–52, in accord with our theory of a regular prytany calendar, we proposed the reading π̣[έμπτει̣ καὶ δεκατ]ῆι τῆς πρυτανείας, giving the equation: <Elaphebolion (IX)> 30 = Prytany VIII [15]. The year would be ordinary. Meritt, in order to support his theory of a Metonic cycle which required that the year in question be intercalary (328/7), responded in *Athenian Year* 95–97 that there was scribal error, claiming that the text had to be π[ροτέραι, <πέμπτει καὶ> εἰκοστ]ῆι. He wrote of the scribe, "One may assume that in his imagination he had finished writing πέμπτηι καὶ when he was only at the end of προτέραι, being confused by the similarity in the endings of the words." Meritt never abandoned this restoration and in his latest archon table in *Historia* 26 (1977) 69 marked the year as a proven intercalary one (I*). C. J. Schwenk, *Athens in the Age of Alexander* 274, who adopts Meritt's calendar throughout, writes of the προτέραι: "Neither Wilhelm, Velsen, Kirchner, not <nor> Meritt (*Ath.*) had any qualms about using this designation for Elaphebolion." Parenthetically, it cannot be said too often that if future calendric scholars rely on Meritt's and Woodhead's calendars for restorations in individual years, their work will be destined for the dustbin. Assuming that what Meritt has one scribe, and one stonecutter, and one archon do with the calendar, others might, scores of Meritt's texts of calendar equations can be altered. There is no evidence for protera except in Skirophorion (XII).

I turn to two of the equations with protera in which these dates had been preceded by thirteen normal months of 30- and 29-day length, and both are of intercalary years with 32-day prytanies.

IG II[2] 495: 303/2 B.C.

> Σκιροφοριῶνος ἕνηι κ[αὶ νέ]αι προτέραι, μιᾶι καὶ τριακοστ[ῆι]
>
> Skirophorion (XII) ultimo[1] = Prytany X 31 = 383 day of the year

Meritt, *Year* 192: 271/0 B.C.[10]

> [Σκιροφο]ριῶνος ἕνει καὶ [ν|έαι προτέραι, μιᾶι καὶ τριακο]στεῖ
>
> Skirophorion (XII) ultimo[1] = Prytany X 31 = 383rd day of the year

In Mikalson's list of eleven "Restorations to give Skirophorion 30," other examples might be cited, but the four above suffice to add to the testimonia that the Athenians used a lunar calendar. For

[10]The many possibilities for the calendar of this year were presented in "The Name of the Game is Restoration," *CSCA* 3 (1970) 208–211. I quoted above p. 54. testimonia which proved that Meritt had misread the name of the archon in an inscription he assigned to this year.

example, we have in *IG* II² 330 a stone which gives three fragmentary decrees (stoichedon) in which the names of the months are all omitted and the numbers of the prytanies are restored. Kirchner assigns the third decree to the tenth prytany of the intercalary year 335/4, giving the text ἕνει καὶ νέαι, ἑβδόμη[ι καὶ τριακοστῆι]. This gives the equation:

<Skirophorion> ultimo = Prytany [X] 37 = 383rd day of the year[11]

The use of προτέρα, found only for Skirophorion (XII) ultimo[1], is in accord with a lunar calendar. A prytany date equated with this day indicates the total number of days in the year before the additional day was added, apparently as a result of observation.

3. Calendar of Leap Days with Alternating Months. A modern theory of an undocumented and unparalleled Athenian calendar of leap years, but otherwise with alternating full and hollow months, is attributed to Meton for which Geminos is claimed as the "indirect" source. This calendar was reviewed at length in "Ancient Athenian Calendars on Stone," *UCPCA* IV 4 (1963) 316–329 and *Choiseul Marble* 54 ff. Dinsmoor, in his *Archons of Athens*, citing in particular the 19-year astronomical calendar of Euktemon cited in Geminos 8.52 ff., evolved a theory that Meton reformed the Athenian festival calendar.[12] The reform, he

[11]Meritt in *Year* 12 accepts Kirchner's restorations for the month and prytany, but claims a day was added in the month Skirophorion. Now, he has claimed in his same volume and throughout his studies, being followed by Woodhead, that Skirophorion was sacrosanct, the calendar being in order by Thargelion (XI) ultimo. In the same publication, he can switch from an untampered Skirophorion to a tampered one.

[12]Geminos 8.52 is an account of the Great Year, or enneadecaeteris, of Euktemon, Philippos, and Kallippos, a calendar in which every Day 64 was omitted. O. Neugebauer, *HAMA* 2.622, writes, "Again on the basis of the relation (1) Meton in Athens introduced a 19-year cycle, beginning at the summer solstice on Skirophorion 13 of the year of the archon Apseudes (–

claimed, was a calendar of regular alternation of full and hollow months broken in periods of 15 and 17 months alternately with two full months (p. 316). This rigid sequence of months from 432 – 109 B.C. was charted on his pp. 424–440 and followed extensively by Agora epigraphists. For the earlier period, Dinsmoor assumed cycles, which nonetheless provided true astronomical dates, as exemplified by his alleged determination of the laying out of the axis of the Parthenon in August 488 B.C.;[13] see "Julian Dates and Greek Calendars," *CP* 42 (1947) 235–243.

In his first publication, *The Athenian Calendar in the Fifth Century* (1928), Meritt tacitly assumed that the civil (i.e. festival) calendar was limited to years of 354, 355, and 384 days; see p. 115. He nowhere offers any explanation or quotes any source. In his second work, *Athenian Financial Documents of the Fifth Century* (1932), he adopted Dinsmoor's sequence of full and hollow months "for all practical purposes" (p. 151). As we have seen, Dinsmoor hypothesized leap days every 15 and 17 months alternately in a rigid pattern. In his *Athenian Year* (1961), Meritt, citing Geminos 8.52, said the passage was authority for an Athenian calendar of alternating full and hollow months, although he never applies such a progressively distorted calendar in his theoretical reconstructions of individual years. Rather he hypothesized without any authority a calendar of rigid alternating months but with irregular leap days. He wrote (p. 34), "From time to time an observation of the new crescent moon (or of any lunar phase) can

---

431 June 27). The date Skirophorion 13 shows that the civil lunar month did not begin near the solstice, a fact that has caused much discussion concerning the "beginning" of Meton's cycle. This problem vanishes, however, when one realizes that Meton did not attempt to introduce a new lunar calendar but intended to establish a definite starting point in the solar year for the construction of parapegmata."

[13]See W. B. Dinsmoor, *The Athenian Archon List in the Light of Recent Discoveries* (New York 1939) 206.

have been the occasion for adding an extra day." Again (p. 35), "We are not told how the Athenians managed the occasional necessary extra day, but they may well have added it, and reversed the alternation, whenever it seemed desirable to do so." In *TAPA* 95 (1964) 241, he wrote that his calendar of the leap day had "no hit or miss determining of the first day of the month by observation of the lunar crescent." In *AE* 1968. 77, he affirmed, "each two month period consisting of 59 days, but sometimes an extra day had to be added to keep the months closer in time with the moon or, at the end of the year, to come out even with the last prytany."

When a scholar sets forth a theory which controls a major portion of his research, in this case the pattern of months within each archon's year, one expects documentation.[14] One looks for a collection of leap days to substantiate his theory. In addition, one would like to be shown that Dinsmoor's rival theory of regular leap days is incorrect, particularly since Meritt had used it for his fifth-century calendar. As there is no literary evidence for either theory, one can only adduce epigraphical evidence in the form of charts for the months of each archon's year. Meritt offered no collection of his leap days. Rather he inserts days into his charts of individual years willy-nilly. No scholar believes that the Athenians had a calendar of strictly alternating months, which over the course of years (**n** years × 0.36707, Heath) would have sacrifices and festivals grossly out-of-step with the moon. The matter of the leap day, if, when, and why it was made, becomes of fundamental importance. Since Meritt asserts that Geminos is the literary source for his calendar, we turn to this author. Geminos' only κατὰ πόλιν calendar is one of 354 days with months ἐναλλὰξ

[14]Woodhead in *Agora* XVI, twenty years after Meritt's publication of his last archon table, reproduces Meritt's calculations for each archon's years without a word in defense of the theory.

πλήρεις καὶ κοῖλοι. Synchronization with the moon would be impossible.

Geminos. The work attributed to Geminos of Rhodes, an *Eisagoge*, or Introduction to (celestial) Phenomena, is a treatise which summarizes the contemporary knowledge of elementary astronomy. Generally dated within seventy years of the beginning of our era, Manitius, the 1898 Teubner editor, believed (pp. 246–248) the work is an abbreviation in the sixth century of an original by Geminos. He based his opinion on the errors and contradictions in the text; compare p. x n. 1 of the Budé edition of Germaine Aujac (1975) and Tittle, *RE* s.v. Geminos 1 (1910) 1038.32–39. Folkerts, in *Der neue Pauly* 4 (1998) 900–901, gives a brief summary by chapters. Geminos treats the zodiac, the motion of the sun, the constellations, the celestial sphere, days and nights, the risings and settings of the zodiacal signs, luni-solar periods in various cycles (of two years, the octaëteris, 16-year and 160-year cycles, the 19-year cycle of Euktemon, and the cycle of Kallippos [76 years]), phases of the moon, eclipses, star phases, terrestrial zones, and geographical places, and the foolishness of making weather predictions by the stars. Probably the most detailed treatment is by O. Neugebauer, *A History of Ancient Mathematical Astronomy* 2 (1975), starting with p. 578 and continuing for a hundred pages throughout his treatment of Greek astronomy. Our concern is primarily with chapter 8 and with the παράπηγμα following chapter 18 with the heading: Χρόνοι τῶν ζῳδίων ἐν οἷς ἕκαστον αὐτῶν ὁ ἥλιος διαπορεύεται.

The parapegma lists under the headings of the zodiac the dates of astronomical phenomena and weather indications collected from those of various astrologoi. It is dated three centuries earlier than Geminos and is pre-Hipparchian. Six authorities for zodiacal parapegmata are quoted in the Geminos passage, the earliest being

Meton and Euktemon and the latest Dositheos.[15] In *BCH* 85 (1961) 31 ff., Van der Waerden and I collected the references to Euktemon's parapegma with an English translation, retaining the zodiacal dates, but calculating Julian dates.

Stone fragments of parapegmata have been found; see Rehm *RE* s.v. (1974) 1295–1366 and Neugebauer, *HAMA* 2.589. In 1902, during the excavation of the theater at Miletos conducted by the German archaeologist Wiegand, four marble fragments were found that were recognized as parts of two parapegmata. A fifth had been found earlier. One parapegma is dated 109/8, the other is slightly earlier. James Evans, *The History and Practice of Ancient Astronomy* (Oxford 1998) 199 ff., has given an excellent study of the parapegmata with sketches of one of the Miletus parapegmas and of a papyrus parapegma from Greek Egypt dated abut 300 B.C. He gives English translations of both as well as of extracts from Geminos and Ptolemy. Neugebauer, *HAMA* 2.587 n. 4, translates lines from the earlier Milesian parapegma. The holes, he explains, represent days without predictions.

Diodoros refers to Meton's 19-year cycle in 12.36 (Loeb tr.):

> In Athens Meton, the son of Pausanias, who had won fame for his study of the stars, revealed to the public his nineteen-year cycle, as it is called, the beginning of which he fixed on the thirteenth day of the Athenian month of Scirophorion. In this number of years the stars accomplish their return to the same place in the heavens and conclude, as it were, the circuit of what may be called a Great year; consequently it is called by some the Year of Meton. And we find that this man was astonishingly fortunate in this prediction which he published; for the stars complete both their movement and the effects they produce in accordance with his reckoning. Consequently, even down to our own day, the larger number of the Greeks use the nineteen-year cycle and are not cheated of the truth.

[15]In Book 2 of Ptolemy's *Phaseis*, twelve are listed with the latest being Caesar. In "Geminos," Meton is mentioned only once, under Cancer.

The reference is clearly not to lunar parapegmata marking the days of festival calendars, but to solar parapegmata according to zodiac signs essential for farmers, seamen, and the like. Cf. O. Neugebauer, *HAMA* 2 p. 617:

> The well-attested lack of influence of calendaric cycles on the civil calendar on the one hand and the equally well-known relationship of cycles to the parapegmata leads to the conclusion that the primary purpose of the known cycles from the 5th and 4th centuries was the construction of reliable parapegmata, not the reform of civil calendars. When Diodorus says that the Metonic cycle was still in use in his time (i.e. around 50 B.C.) "with the majority of Greeks" he obviously did not expect his readers to think of the civil calendars and he refers explicitly to Meton's success in predicting the "significant" days for the weather prediction from the stars.

The Diodoros passage with others are collected by Rehm, *RE* s.v. Parapegma (1949) 1298–1299; δεξάμενοι τοίνυν οἱ μετὰ Μέτωνα ἀστρονόμοι πίνακας ἐν ταῖς πόλεσιν ἔθηκαν περὶ τοῦ ἡλίου περιφορῶν τῶν ἐννεακαιδεκαετηρίδων (Schol. Aratos). Cf. Aelian *VH* 10.7 (ἀνέστησε στήλας καὶ τὰς ἡλίου τροπὰς κατεγράψατο). These parapegmata, including that of Meton, gave what Evans calls "star phases", as indicated by the term ἀστρόνομοι, and must have been very popular, which may explain why Geminos inveighs against them.[16] They had nothing to do with lunar parapegmata, only one fragment of which has been found, that in the excavations of the Pompeion: Brueckner, *Ath. Mitt.* 56 (1931) 23–24, dated in the third century B.C.[17]

[16]See chapter 13 with Notes complémentaires of Aujac.

[17]I formerly wrote that there were two, but *IG* II^2 2782, is a republication, as a student in Toronto caused me to see.

In his *Athenian Year* (1961) 25 n. 12 and 34–35, Meritt cited the 8.52 passage where Geminos is contrasting the Euktemon cycle with an earlier octaëteris. He claimed by "indirection" the passage provided evidence for an Athenian festival calendar of alternating months. He cited no other ancient authority nor analyzed the passage. After my critique of his interpretation in *Stone* 316–329, Meritt indicated his error in an oblique fashion. P. Clement, who worked in an adjacent office and has never otherwise written on calendric matters, published a review of Meritt's *Year* in which he wrote the following (*AJA* 69 [1965] 192):

> Meritt's assertion that Geminus provides "indirect evidence" for Athenian use of "a scheme of alternating months" of 30 days and of 29 days is of course correct. The crucial item of evidence lies in paragraph 3 of Chapter VIII of the Εἰσάωγή: <Greek text> Ἔστι δὲ ὁ μὲν ἀκριβὴς μηνιαῖος χρόνος, καθάπερ εἴρηται, ἡμερῶν κθ´ ϛ´´ λγ^{ου}, ὁ δὲ πρὸς τὴν πολιτικὴν ἀγωγὴν ὁλοσχερέστερον λαμβανόμενος ἡμερῶν κθ´ ϛ´´, ὥστε τὴν δίμηνον γενέσθαι ἡμερῶν νθ´. ὅθεν διὰ ταύτην τὴν αἰτίαν οἱ κατὰ πόλιν μῆνες ἐναλλὰξ ἄγονται πλήρεις καὶ κοῖλοι διὰ τὸ τὴν <κατὰ> σελήνην δίμηνον ἡμερῶν εἶναι νθ´.
>
> By some mischance citation of this particular passage has fallen from those parts of *The Athenian Year* in which reference to Geminus is made. This is a pity, for, if Pritchett had read the passage, he could have been spared needless writing.

I give Aujac's translation:

> La durée exacte du mois lunaire est donc, comme je viens de le dire, de 29 jours 1/2 1/33; mais dans la vie courante on utilise un chiffre arrondi à 29 jours 1/2, si bien que deux mois valent 59 jours. C'est ce qui explique que les mois civils soient alternativement pleins et creux: les deux mois lunaires y valent 59 jours. Il en résulte que l'année lunaire se compose de 354 jours: en multipliant par 12 les 29 jours 1/2 du mois, on obtient pour l'année lunaire 354 jours.

Geminos, in writing of the octaëteris, which, according to him, was the first cycle that the ancient astrologoi constructed, has a similar passage in 8.34–35, on which T. L. Heath, *Aristarchus of Samos* 287, comments and translates:

> On the alternation of 'full' and 'hollow' months an apparently interpolated passage in Geminus says:
>
> 'The moon-year has 354 days. Consequently they took the lunar month to be 29 1/2 days and the double month to be 59 days. Hence it is that they have hollow and full months alternately, namely because the two-months period according to the moon is 59 days. Therefore there are in the year six full and six hollow months. This then is the reason why they make the months full and hollow alternately.'

Geminos follows this by saying (Heath's tr.):

> 'Observation having speedily proved this procedure to be inconsistent with the true facts, inasmuch as the days and the months did not agree with the moon nor the years keep pace with the sun, they sought for a period which should, as regards the years, agree with the sun, etc.'

The astrologoi in publishing their zodiacal parapegmata sought periods of months which would agree with the sun. Such periods had nothing to do with lunar calendars of Greek city-states.

One will look in vain in the numerous writings of Meritt and his followers for a defense of a calendar of 354 days with alternating months used through the centuries which progressively fell out of step with the moon. Meritt, himself, returned to the subject in *TAPA* 95 (1964) 241, still affirming a calendar of alternating months and claiming that Geminos ruled out a lunar calendar, citing the 8.3 passage of Clement and 8.7.

Aujac, in her Budé edition, has conveniently given sub-titles for sections of Geminos. In 8.6–15, she gives the subject as Le calendrier grec, followed by Le calendrier égyptien. Heath's English translation (p. 285) was given at the beginning of this work. Geminos attributes to ἅπαντες οἱ Ἕλληνες a calendar which he repeatedly says is κατὰ σελήνην (τὰς δὲ ἡμέρας καὶ τοὺς μῆνας τῇ σελήνῃ). He specifies that the calendar was according to the phases of the moon (τοῖς τῆς σελήνης φωτισμοῖς). He explains the effect as having sacrifices fall at their designated time. He quotes a long passage from Aratos (733 ff.) which has the progression of days of the month ἐκ τῶν φωτισμῶν τῆς σελήνης. Geminos devotes a separate chapter (9.1–16) to Περὶ σελήνης φωτισμῶν. The κατὰ σελήνην calendar which Geminos attributes to all the Greek states is a lunar calendar, corresponding exactly to the κατὰ θεόν calendar of the Athenians and Boiotians.

As to Meritt's 'overlooked' 8.3 passage, in 8.1–5 Geminos has a brief section, Le mois lunaire, in which he uses the phrase κατὰ σελήνην in referring to a sequence of full and hollow months: ... οἱ κατὰ πόλιν μῆνες ἐναλλὰξ ἄγονται πλήρεις καὶ κοῖλοι διὰ τὸ τὴν <κατὰ> σελήνην δίμηνον ἡμερῶν εἶναι νθ΄. Chronologists today who believe in a calendar κατὰ σελήνην often find it convenient for any two given months to assume a sequence of 30 + 29, which of course is very common, without explaining other possibilities (1±) in a lunar calendar. Geminos is using popular language, as noted in *Stone* 328 and *Choiseul Marble* 56. Heath (*Aristarchus* p. 285) writes, "Alongside the 'full' and 'hollow' months of the calendar, popular parlance invented a month of 30 days, as being convenient to reckon with." See Sontheimer, *RE* s.v. Monat (1933) 45.46–51, and Ginzel (*Handbuch* 2.366–369). Cf. Mommsen, "Monate im Sprachgebrauch," *Chronologie* 48–51. O. Neugebauer, *HAMA* 2 p. 619, in computing early cycles, uses the phrase, "The commonly accepted round value of 29 1/2 days for the lunar month". Webster's *Unabridged Dictionary* s.v. Month states, "In popular use, a period of four weeks or of thirty days is often called a month." In 8.3 as in 8.6–15, the calendar Geminos attributes to the Greek states was κατὰ σελήνην.

In *Calendars of Athens* p. 5, Neugebauer wrote,

> The months of a lunar calendar form rather irregular sequences of intervals of 29 and 30 days whose arrangement cannot be predicted without a highly refined astronomical theory. This stage of development was reached only in the third century before Christ in Mesopotamia. Without theoretical astronomy, actual observation alone can determine the beginning of the months. In a lunar calendar, sequences amounting to four consecutive 30-day months and to three 29-day months are possible, though rare. Pairs of 30-day months or 29-day months, however, are very common. This may be illustrated by an arbitrary example from a Babylonian ephemeris of 104/3 B.C.

In conclusion, the evidence of Geminos is that the Greek states used a calendar κατὰ σελήνην, "according to the goddess Selene". The astrologoi, apparently publishing treatises much longer than the extracts in Geminos, used artificial months for their zodiacal parapegmata.

Epigraphical Evidence. I turn from the literary evidence for Meritt's theory about the sequence of months to the epigraphical, focusing on the leap days which Meritt claimed interrupted his sequence of full and hollow months. Meritt put forth his theory of irregular leap days in 1961 in his *Athenian Year* without offering a collection of such days to illustrate his theory on which it was based. Table 5 is a collection of his 30-day months where he has two in direct sequence, thus interrupting his schematic calendar of alternating full and hollow months. With Meritt it is always important to examine the evidence on which his theories are originally based and which he subsequently claims he has proved. Woodhead in *Agora* XVI always follows him blindly. I limit myself to the examples where he offers a chart of the sequence of months in individual archon years.

TABLE 5

MONTHS WITH LEAP DAYS IN MERITT'S *ATHENIAN YEAR*

PAGE	YEAR	MONTH	
78	337/6	12	
79	336/5	13	
82	335/4	3	
83	334/3	9	
85	333/2	10	
91	331/0	11	
95	329/8	1	(following Skirophorion with 30 days in preceding year)
97	328/7	10	
99	327/6	6	
104	325/4	1	(following Skirophorion 30)
106	324/3	7	
109	323/2	6	
112	322/1	13	
120	320/19	4	
124	319/8	11	
127	318/7	5	
129	314/3	9	
137	247/6	9	
146	252/1	13	
150	244/3	11	
156	188/7	3	
159	178/7	2	
159	173/2	2	
163	160/59	7	
177	307/6	13	

In the twenty years from 337/6 through 318/7, sixteen had leap days. Only Month 8 did not have a leap day. If the twenty-year period is a ample of observations over the centuries, Meritt has attributed to the Athenians a highly schematic calendar which cannot be synchronized with the moon. A mathematician might figure out where his noumenias might fall. The observers Meritt hypothesizes ordered his calendar were far inferior to those who determined the lengths of months in attested historical lunar calendars, although he assures us that they made "no hit or miss determining of the first day of the month by observation of the lunar crescent".[18] Moreover, of the twenty years, Meritt allows embolimoi/exairesimoi days in only one year, and that for two days only. In short for him the festival or kat' archonta calendar was the determining element in Athenian time-reckoning in this particular 20-year period. Hence, we find that for Meritt there are no two years with the same pattern for the length of prytanies, which most agree, I believe, were determined in advance. Indeed, the main tenet in the *Athenian Year* is the argument that neither the Aristotelian nor any fixed pattern applied in any period. Allowing that what one archon did with his calendar another might, and consulting Table 1 of triple dates (p. 3) or Table 4 of Meritt's embolimoi days (p. 21) in chapter 1, one can produce quite different charts for all of Meritt's and Woodhead's years.[19] Entirely

[18]In the same period, Dinsmoor's perfect theoretical calendar (*Archons of Athens* pp. 428–430), followed by Meritt in his early studies, has 13 leap days. In a century, Meritt's calendar would be off by 15 days.

[19]In *AE* 1968. 87, Meritt writes, "In Athens with all her literature, and with all the epigraphical and archaeological evidence now available, it is astonishing (were observation the rule) that nothing is said about observing the lunar crescent ...". What is astonishing is that a scholar would write such a phrase and not feel the slightest compunction to offer any documentation or tabulation for an unparalleled calendar of irregular leap days that required more skill by observers than for well-attested κατὰ σελήνην calendars. On one page, Meritt writes that the Athenians could not

apart from debatable readings and restorations, the epigraphical evidence Meritt adduces does not confirm his theory of alternating months with leap days, which also holds for the literary evidence studied above.

One caveat may be given for the future. The evidence for a true lunar calendar is so strong that anyone who wishes to revive the theory of leap days or claim that the Athenian year was limited to 354, 355, and 384 days, rather than 354 ±1 and 384 ±1 days as in all lunar calendars, must prove that the latter was not possible, and that starting from a clean slate. References to the charts and asterisks of Meritt and Woodhead will be based on false premises.

Other scholars have accepted Meritt's thesis of the Athenian moon-year and thrown darts at the theories put forth in *Calendars of Athens* and later articles. In "The Omitted Day in the Athenian Hollow Month," *ZPE* 41 (1981) 107–124, J. A. Walsh followed Meritt's game of terminology, but differed in claiming that the length of each month was determined by observation on the 20th day. I responded to his theories in "The Calendar of the Gibbous Moon," *ZPE* 49 (1982) 243–266, to which a brief addendum may be offered. He accepts Meritt's basic theory that the Athenian calendar was unique,[20] differing from all other Greek calen-

---

handle observation at the end of the month; on another, observation by looking "at any lunar phase" is the basis for his leap days. The two major premises for his theory are false: (1) a virtually untampered festival calendar, which he later contradicted (Table 4), and (2) accurate observation, which Table 5 disproves. In effect Meritt disproved his own theory, which, however, he continued to advocate to the end without a backward glance.

[20]Meritt repeatedly stipulated that only evidence specifically Athenian should be admitted for the Athenian calendar, ignoring the fact that Geminos, his literary authority, used only the general phrase οἱ κατὰ πόλιν μῆνες, but in particular because the Rhodians omitted Day 29, as attested in *IG* XII.1.4. As noted in *Calendars of Athens* 25 n. 76, "Present

dars and lunar calendars in general; for example (p. 113), "the Athenian month is not set up like Greek calendars which we know omitted the month's second to last day."[21] He concedes that calendars with τριακάς as the last day omitted δεύτερα φθίνοντος, but ignores the fact that calendars of other states share common conditions. He attributes his calendar to "Solon's genius", but nowhere postulates that Solon omitted a specific day, rather one omitting a date "some time before the beginning of backward count" (p. 113). "It seems certain that the post-Solonic calendar was set up so that δεκάτη/ἐνάτη φθίνοντος was the omitted date" (p. 116). He tells us on p. 113 that "Solon's genius" produced a calendar in which the length of the month was determined at "the beginning of backward count," whereas the pre-Solonian calendar gave no "advance warning of the month's last day" (p. 112). Walsh gives no hint as to what ancient source could have provided such information of Solon's purpose. Surely, Solon would have omitted a specific date; it makes no sense to claim that he legis-

---

reconstructions of this inscription are based on the assumption of alternating hollow and full months and should be reconsidered." All lunar calendars had stretches of alternating months. Months 5 and 9 were full; 6, 8, and 12 hollow. W. R. Paton (*Inscriptions of Cos* [1891] 328) claimed by his reconstruction of the four columns, "It will be found that the Calendar commenced in line 4 of col. 1, the first month being a month of 31 days." An interesting feature is that the intercalary Panamos B was the thirteenth month, whereas Panamos was the tenth.

[21]On p. 113 of *ZPE* 41 (1981), Walsh wrote that the moon's progress could be checked "when the tips of the moon's crescent are in line ... The εἰκάδες provided a helpful point at which the moon's progress could be checked ... The decision 'full or hollow month' was made before the second of the two days which constitute the εἰκάδες ... Those responsible for the determination on the moon's length may have wishes to have the added advantage of noting the moon's progress on both of the εἰκάδες—the 20th and 21st." Actually in the gibbous phase (meaning more than half and less than the whole) the moon has no tips. In the average moon-month, on Day 20 the moon would rise in the east about 10 p.m.

lated on the assumption that he anticipated a future change in counting centuries later. At the beginning of his article, Walsh, reproducing Meritt's game of terminology, says he is arguing for two separate Athenian calendars which differed for Days 20 and 21. The earlier one he charts as using εἰκάς and δεκάτη φθίνοντος, the later δεκάτη προτέρα and δεκάτη ὑστέρα. He puts the change in "the late 5th century," but gives no clue as to what was wrong with the earlier calendar, why the change was made, or who prompted it. He does not mention the name of Meton, as Meritt has it, apparently because he realized that the calendar attributed to Meton omitted every 64th day and Solon could not have anticipated that. We know that δεκάτη προτέρα occurs on the Choiseul Marble. Now it makes no sense to claim that the Athenians had two different calendars controlling what day was omitted and hence how inscriptions are to be restored unless the date of change was hard and fast. They clearly did not use two different calendars simultaneously. For example, after Donnay and others pointed out that δεκάτη ὑστέρα (Day 21), as well as δεκάτη προτέρα, occurs several times in the Erchia fasti and was not an omitted day, Walsh, *ZPE* 41 (1981) 122, claimed this date "was, in fact, never omitted and (was) present in every month." Yet εἰκάς is found in Athenian sources after 407/6 B.C., as Meritt realized. For Walsh it was used only in the early period. It would be a service if calendric scholars ceased the game of terminology introduced by Meritt. Walsh continues that if the Athenians used a date with φθίνοντος, it must be understood as that specific day in the count before the month's end. His argument is twofold: (1) Each day in the backward count was interpreted literally only by the Athenians. 'Ογδόη θφίνοντος must be the eighth day before the month's end (p. 114), whereas other states had calendars in which that day might be the eighth (full) or seventh (hollow). (2) "Athenian economic life ... could not have been governed by a calendric system which provided no advance notice of the month's last day" (p. 115). But other Greek states, and surely the Athenian, lived with a calendar in which no ambiguity existed until Day 29. If ambiguity is to be the criterion for the omitted day, it is obviously better to have 28 unambiguous days than 19. Waiving the fact that much of the Athenian administrative and economic life was governed by the prytany calendar, we turn to the question of who needed this "advance notice" of the month's last day, particularly the administrators of what festival, since we are dealing with the festival calendar. "Festival calendar" and "prytany calendar" are not terms found in Walsh's article. Certainly, the administrators of the Greater Panathenaia, apparently of four days,[22] in the pre-20 days of Hekatombaion would not welcome any uncertainty about the exact number of days for preparations of their processions, games, etc. The same would be true of any festival in the mid-twenties, particularly for those coming from Rhamnous and Sounion, although we do not know the dates of many festivals. Walsh's crowning example, and his only specific one, is not an advance notice for a festival, but for a gathering of Plataians. In Lysias 23 *Against Pankleon* 6, the speaker says (modified Loeb), "They told me that I should get the most definite information if I went εἰς τὸν χλωρὸν τύρον ἐν τῇ ἕνῃ καὶ νέᾳ for on that day in each month the Plataeans collect there." I quote Walsh's commentary at some length (pp. 122–123) because it illustrates the absurdity of the "advance notice" theory.

> It is clear from this text that the fresh cheese market was held on one day in every month (ἕνη καὶ νέα). Those who wished to sell cheese (particularly those who sold fresh cheese) must have known in advance when this day was to be. It is absurd to believe that the decision on the month's length was reached only on the 28th (pace

[22]See Pritchett in ΦΙΛΙΑ ΕΠΗ ΕΙΣ ΓΕΩΡΓΙΟΝ Ε. ΜΥΛΩΝΑΝ 2 (1987) 181.

> Pritchett) and that messengers were sent out immediately to notify those interested in selling and buying cheese that the market would be held tomorrow (in the case of a hollow month). It is no less absurd to believe that another day's notice (in the case of a full month) would have been sufficient. These reliable cheese merchants could not have been expected to prepare and transport their wares with even as much as 24 hours notice. It must be remembered that those who sold cheese were country folk, many living a distance from the city and in the hills, as they do today. The regularity with which this market was held suggests that they showed up on the same day of every month (τοῦ μηνὸς ἑκάστου). To do so they must have had a workable calendar. Such regularity could have been provided only by a calendar which gave a number of days warning before the end of the month. This calendar is only possible of the date at the beginning of the backward count δεκάτη/ἐνάτη φθίνοντος) was omitted. ... It is unlikely that administrators, any more than merchants of fresh cheese, would have found the short notice afforded by Pritchett's calendar sufficient for the smooth operation of the Athenian state.

What is certainly not clear is that Athenians could buy fresh cheese only on the last day.

As to administrators of a state not being able to work with a calendar in which Days 1–28 were determined in advance, it is admitted that those of all other Greek states did so. There is no evidence of Athenian festivals or sacrifices on Day 29. One expects Walsh to document what circumstances were peculiar to Athenian administrators of festivals on the last day.

Lysias does not say that Day ultimo was the only one when the Athenians could buy cheese. There is no implication that cheese was sold only on the last day. Walsh has misread the passage, which says no more than that the Plataians collected εἰς τὸν χλωρὸν τύρον on the last day. What was the χλωρὸς τύρος? Guillaume in DS, *Dictionnaire des Antiquités* 1 col. 151A s.v. Agora writes,

> Pour que les acheteurs pussent se diriger facilement dans un pareil dédale, il y avait des quartiers spécialement affectés à chaque genre de produits, quartiers que la nouvelle comédie désignait sous le nom de κύκλοι. Chaque quartier portait, comme nom spécial, le nom de la marchandise qu'on était certain d'y rencontrer. On allait au poisson (εἰς τοὔψον), au fromage blanc (εἰς τὸν χλωρὸν τύρον), aux pots (εἰς τὰς χύτρας), au vin (εἰς τὸν οἶνον), etc. Les marchands étaient installés, les uns en plein air, d'autres sous de modestes tentes d'étoffe, d'autres dans de petites boutiques mobiles, faites de clayonnages ou de roseaux. Quelques-uns avaient de véritables magasins. Des ateliers de toutes sortes étaient établis dans le voisinage et finissaient par se confondre avec l'agora.

See also Szanto, *RE* s.v. Agora (1893) 881, for a larger list of special markets. That such markets were one-day affairs is ridiculous, and Lysias does not say the χλωρὸς τύρος was, only that the Plataians collected there on the last day of the month. The phrase χλωρὸς τύρος is well attested; see Pollux VII.175 (X.130), and particularly Aristophanes *Frogs* 559, with Sommerstein's commentary adding other references. New, soft cheese was kept in baskets (τάλαρος) of wicker-work in which the cheese was placed so as to let the whey run off and nothing remained but the dried cheese. Kroll, *RE* s.v. Käse (1919) 1489–1496, gives a good discussion including our phrase on col. 1492.[23] Kroll (1493.22–25), citing Aristophanes *Eq.* 480, thinks the cheese of the Lysias pas

[23]Cougny, DS *Dictionnaire* s.v. Caseus, vol. 1 p. 932, explains the process with references: "Le poëte Prudence a décrit en vers élégants cette délicate opération. Le fromage se place ensuite dans un panier qu'on suspend en plein air et qui s'appelle κρεμάθρα, κρεμαστήρ, etc. Tant que le fromage est dans les éclisses (τυρὸς ἐκ ταλάρων) il est dit fromage vert (χλωρός, *viridis*, vulg. ἁπαλός), c'est-à-dire frais, blanc ou mou, ou ἀνθῆνας (ἀνθείνας, *floridus*). Dans cet état, on le trouvait délicieux, nourrissant et d'une digestion facile."

sage came from Boiotia, but the Athenians imported cheese from distant places, and doubtless made their own. That a wine-merchant or a cheese-merchant required "advance notice" is Walsh's creation. Cheese and wine, as staples, must have been purchased on many days of the month, especially on the noumenia which Nilsson shows was a great market day. There were several stoas in Attika, see Aristophanes *Ach.* 548, *Eccl.* 686.

A fatal objection to Walsh's calendar is that he would have his Solonian astrologoi, after observing the moon about 10 P.M. on Day 20 of Boedromion, tell the officials of the Eleusinian Mysteries, if the month was to be hollow, that they had only a few hours to cancel one of the principal days of the festival (Day 21). We know from two of our earliest inscriptions (*IG* I³ 5 and 6) that great attention was paid to the Eleusinia as a great early festival. In *ZPE* 35 (1979) 145–151, "The Omitted Day in Athens and the Mysteries," Meritt attempted to revise the calendar of the Mysteries in Boedromion to show that ἐνάτη μετ' εἰκάδας was the omitted day. I responded in "The Last Day of the Eleusinian Festival," *ZPE* 41 (1981) 145–148, to show that Boedromion 23 was the last day, following Mommsen and Mikalson, and that δευτέρα μετ' εἰκάδας was necessarily the omitted day. Walsh shows no awareness that what he terms the "Athenian calendar" was a festival one. He mentions no festival in the entire article. It was quite a feat to advance a completely undocumented theory on the Athenian calendar without mentioning festivals or prytanies. For most of the article, he parrots Meritt.

Cynthia J Schwenk, *Athens in the Age of Alexander* (1985) 151–157, in republishing Meritt's text of *IG* II² 339A,[24] adopted

[24]In her commentary on *IG* II² 339A, Schwenk devotes pages 152–157 to the calendar including tables, reaffirming Meritt's "guaranteed" restoration of this small fragment with neither right nor left margin in which the craftsmanship of the mason was poor. I have commented on the possibilities in *Calendars of Athens* 46–48; *Stone* 276–277, 282, 378, with

his restorations based on the letters νος εκτ/ιδε, which appears to have convinced her of Meritt's omitted days thesis. She wrote in part in support of Walsh:

> It has never been questioned that Metageitnion of this year was hollow with 29 days. Pritchett-Neugebauer, Pritchett (in his later article) and Mikalson, contend that the omitted day for a hollow month was δευτέρα φθίνοντος, the last day but one. This is based on literary evidence—chiefly scholia—and the assumption that the Athenians determined their festival calendar by actual monthly observation of the new moon and thus would not know before the end of the month whether it would be hollow or full. Meritt, on the other hand, following different scholia and the assumption that the Athenians knew the length of the month at least by the 20th, believes that the omitted day was more likely the 21st or 22nd; in any case, *not* δευτέρα φθίνοντος. J. A. Walsh, in his article, "The Omitted Date in the Athenian Hollow Month," supports Meritt's opinion. Indeed, Walsh's examination of previous scholarship, and his investigation into the significance not only of the Solonian calendar reforms but also the term used to designate a 29-day month, "hollow," point to the definitive conclusion "that δεκάτη/ἐνάτη φθίνοντος was the omitted date in the Athenian hollow month," (124) thus rendering any subsequent argument concerning this specific calendar "problem" pointless.

As to a "definitive conclusion" about the word κοῖλος, the three scholars can be hoist with their own petard. Meritt (*Mnem.* 30 [1977] 241), duplicated by a lengthy note in Walsh (*ZPE* 41 [1981] 115), advances as a crowning argument the theory that κοῖλος could not describe a month that omits the next to last day.

---

photograph in pl. 21; and above p. 37. A scholar, unacquainted with the complexity of the problem, finds one text with square brackets and generalizes that it affords conclusive evidence for the calendar. Any scholar who wishes to contribute to the problem of the festival calendar starts with the kata theon dates, not texts in square brackets.

The *OLD* p. 291, observes that cavus in Latin was used "of the crescent or waning moon", citing Pliny *NH* 8.215 (Loeb) "are depressed by the moon waning (luna cava) and worship the new moon". Frontinus *Str.* 1.5.1 uses the word for an object "having a crescent-shape (cavae lunae)". Lewis, *Latin Dictionary* 307, observes, "Menses, which have only 30 days (opp. menses pleni, which had 31 days)". Cavus was most often used for any object having a hole. So in Greek according to LSJ, the word is used for the hold of a ship, "of places, lying in a hollow or forming a hollow", the bed of a river, concave (often). The substantive means hollow, empty; κοιλότης hollowness, concavity. The word for "hollow, concave" in Modern Greek is κοῖλος. The hollowest time of the moon in crescent shape is not Day 20 in the gibbous phase, but at the end of the month: δευτέρα φθίνοντος (Day 29). Of the arguments used to support the theory of the omission of Day 22, this seems the strangest.[25] Meritt advanced the idea after he had withdrawn his statement that he had proved that by forward count the omitted day was the 29th.

Conclusions to Chapter IV

The primary support for the theory of a lunar calendar is the fact that κατὰ θεόν = κατὰ σελήνην = "according to the goddess Selene". Calendars so designated were lunar.

The last day of the year (Skirophorion ἕνη καὶ νέα) might be the 353rd or 383rd day, which also speaks for a lunar calendar.

So strong is the evidence that the Athenians had a lunar calendar of 354 ±1 and 384 ±1 days that any scholar who wishes to revive the theory that years were limited to 354, 355, or 384 days should

[25]Cf. *ZPE* 49 (1982) 255 n. 35.

start by proving that the ±1 years were impossible, starting with a clean slate and disregarding charts which are only exempli gratia.

No scholar subscribes to an Athenian festival calendar of strictly alternating full and hollow months.

Geminos, the sole literary source cited for such a calendar, attributed to the Greek states a κατὰ σελήνην calendar.

The astrologoi in Geminos used artificial months for their zodiacal parapegmata.

Meritt's epigraphical evidence for his leap day festival calendar results in a grossly distorted one.

The theory of an observation festival calendar with lengths of months determined on Day 20 to give advance notice of Day ultimo (only) was poorly conceived and lacks any credibility.

The word κοῖλος applies to the shape of the crescent at the end of the month.

V

PRYTANY CALENDAR

The length of prytanies calculated from epigraphical evidence has led all scholars to posit that the prytany calendar, at least down to the last decade of the fifth century, was different from that in the Aristotelian period by which time Prytany I 1 was equated with Hekatombaion (I) 1.[1] I treat the two periods separately.

Fourth and Later Centuries. In the first half of the fourth century, Athenian prescripts might give the number of the prytanizing phyle and day of the prytany, but never the month or day of the month. Any tabulation of the official copies in the archival records must have been by the prytany calendar.[2] Limiting myself to *IG* II², the first occurrence of the day of the month is found in *IG* II² 229 (341/0); the name of the month is omitted.[3]

After World War II, Neugebauer and I inherited a situation in which the festival calendar was schematic, the determining element in interpreting calendar equations. Various lengths were assigned to the prytanies. Meritt stated in *The Athenian Calendar*, Preface, p. vii:

> One fundamental assumption has been made in the calculations which involve maxima and minima, viz., that the length of a pry-

[1]Meritt disagrees with this equation in selected years, where I disagree.

[2]The case against archival originals is presented by R. Thomas, *Oral Tradition and Written Record in Classical Athens* (Cambridge 1989) 45 ff. But see Sickinger, "Inscriptions and Archives in Classical Athens," *Historia* 43 (1994) 286–296, with which I am in agreement. See *Greek Archives, Cults, and Topography* (1996) 14–39.

[3]Such omission of the month continues in *IG* II² 330 (bis), 331, 354, 357 (327/6). Cf. A. S. Henry, *Mnem. Suppl.* 49 (1977) 37.

> tany may have been as much as forty days or as little as thirty-three days. No one who has studied the problems of the Athenian calendar, I am sure, will consider these limits too conservative.

And again, on p. 41:

> For the sake of uniformity we shall consider 40 and 33 as possible maximum and minimum numbers of days in a prytany, where the reckoning of such lengths of prytany is necessary, although again we have chosen limits more widely divergent than necessary for the sake of safety in argument.

Dinsmoor, *Archons of Athens* (1931) 329, wrote in similar fashion, 'We cannot employ these irregularities of the administrative <= prytany> calendar as arguments for corresponding irregularities in the astronomical lunar or civil <= festival> calendar, which was unaffected by official procrastination."

There are four possibilities:

1. Both the festival and prytany calendars were fixed and schematic.

2. The festival calendar was regular; the prytany calendar irregular.

3. The festival calendar, being lunar, was irregular, with exairesimoi/embolimoi days; the prytany calendar was regular.

4. Both the festival and prytany calendars were irregular.

No scholar accepts no. 1. No. 2 has been the position of Ferguson, Dinsmoor, and Meritt, the last through 1961 (and often thereafter). Neugebauer and I advocated no. 3. No. 4 becomes the practice of Woodhead in *Agora* XVI (1997) in particular and often Meritt and his other followers.[4]

[4]Cf. *BCH* 81 (1957) 269–301.

No explanation has been offered as to why civil authorities should divide years of 354 ± 1 days and 384 ± 1 days into prytanies in capricious fashion, 33 days in one prytany, 36 in another, and 40 in yet another, with one year rarely agreeing with another.

In *Calendars of Athens* (1947) 34, Neugebauer and I introduced the previously ignored statement of Aristotle *Ath. Pol.* 43.2 that in a year of 354 days the first four prytanies had 36 days and the next six 35. We proposed similar regularity in other periods of a lunar calendar (354 ± 1 and 384 ± 1 days), since Aristotle says the year was κατὰ σελήνην. We referred to similar statements, apparently derived from the *Ath. Pol.*, in Harpokration, Photios, Souda, *Lex. Bekk.*, *Lex. Rhet.*, *Lex. Patm.* Harpokration says, ἐνιαυτὸν δὲ οἱ Ἀθηναῖοι τὸν σεληνιακὸν ἦγον.[5] The interesting feature is that these lexicographers and scholiasts cite no contrary opinion. The tradition is pure.

We failed to cite the scholia to Plato *Laws* 12.935C, which in its details must derive from a different tradition (text of W. C. Greene):

953 c πρυτάνεων.

πρυτάνεις εἰσὶ μέρος τι τῶν φ´ τὸ δέκατον, ν´ ἄνδρες, οἱ διοικοῦντες ἅπαντα τὰ ὑπὸ τῆς βουλῆς πραττόμενα, καὶ ἐπρυτάνευον κατὰ πρυτανείαν ἐκ διαδοχῆς. πρυτανεία δέ ἐστιν ἀριθμός τις ἡμερῶν, ἤτοι λϛ´ ἢ λη´ ἢ λε´, ἅς ἑκάστη φυλὴ πρυτανεύειν λέγεται. ιβ´ μὲν γὰρ οἱ μῆνες, ι´ δέ εἰσι πρυτανεῖαι, καὶ φυλαὶ ι´. καὶ διῄρηνται εἰς ταύτας αἱ ἡμέραι τοῦ ἐνιαυτοῦ· κατὰ γὰρ σελήνην ἄγουσι τοῦτον, ὡς ἑκάστῃ φυλῇ τῶν ι´ ἐπιβάλλειν λε´ ἡμέρας, πλεονάζειν δὲ ὀλίγας. διὸ καὶ τὰς λοιπὰς ἀπέδωκαν οἱ Ἀθηναῖοι ταῖς πρώταις λαχούσαις τέτταρσι φυλαῖς, ἵνα ἐκείνων μὲν ἑκάστη τὰς λϛ´ ἡμέρας πρυτανεύῃ αἱ δὲ λοιπαὶ ἓξ ἀνὰ λε´, πρυτανεῖα δὲ

[5]Some use the phrase κληρούμενοι τὸ δέκατον μέρος. LSJ cites selene as used for a month in Euripides *El.* 1126 (δεκάτῃ σελήνῃ, in the tenth month), *Alc.* 431, *Tr.* 1075, *Hel.* 114.

οὐδετέρως οὐχ εὕρηται· τινὲς δὲ ἀργύριόν τι, ὃ κατατίθεται ὑπὸ τῶν δικαζομένων, καὶ δίδοται δικαστικὸν τοῖς ἑξακισχιλίοις.

In writing "either 36 or 38 or 35" with the extra days being given to the first four prytanies, the 36 and 35 apply to an ordinary year, but the 38 to an intercalary year. The scholion would be completely in accord with our thesis if one assumes that λθ´ (39) had dropped out in line 4. The numeral for 38 requires 39 to complete the context. The value of a Byzantine scholion is hard to appraise, but it applies only to the Aristotelian period before 307/6 and presumably from a sound source. W. C. Greene, *Scholia Platonica* (1938) writes in his Praefatio p. xxviii, after citing earlier literature, "Demonstraverunt scholia Platonica non una aetate nedum ab uno scholiasta esse congesta." All of the literary evidence was in accord with our theory.

Financial Documents. In *Calendars of Athens*, we also assembled evidence in the various periods of the few, but all-important cases where the length of prytanies was determined by epigraphical evidence other than by the game of restoring calendar equations in prescripts. For example, in the Aristotelian period, we cited (p. 36) the evidence of *IG* II^2 1672, where Kirchner on p. 311, bottom of 2nd column, reported that the first and second prytanies in the year 329/8 had 36 days, the fifth and sixth 35; see now W. T. Loomis, *Wages, Welfare Costs and Inflation in Classical Athens* (1998) 279. The same inscription shows, according to Loomis, *Hesperia* 64 (1995) 132–133, that the length of the last prytany in the intercalary year 330/329 was of 38 days. The latter disproves Meritt's chart of prytanies in *Athenian Year* p. 94. Indeed, Meritt mentions *IG* II^2 1672 only once (p. 95) in his career, omitting it from his index in *Year*.

To anticipate our treatment of prytany dates in the next section, in the fifth century, were the Aristotelian pattern would require six prytanies of 37 days followed by four of 36 days, we have in the Erechtheum building documents published as *IG* I[2] 374 and *IG* I[3] 476 similar records of pay. In this document, the sixth prytany (lines 108–112; 59–62) had 37 days and the eighth 36 days: Loomis, *Wages* p. 99. Meritt never mentions this text. For other fifth-century evidence which is not contested, see *Calendars of Athens* 95–96.

For men receiving wages, pay was made per diem by the prytany calendar, and a distinction was made of one day. This evidence alone bespeaks for a schematic calendar with the number of days in each prytany allotted in advance according to the number of prytanies.

Highest Dates. In *Calendars of Athens* Chapter 3, we offered texts (including restorations) for intercalary years of 384 ±1 days in the period of twelve phylai after 307/6 in accord with the principle of even-length prytanies. Each prytany was of exactly 32 days' duration with the exception of the twelfth which would be lengthened or shortened by one day. Thus the ἕνηι καὶ νέαι of the last month in *IG* II[2] 945 fell on the 31st day of the twelfth prytany. Although the designation of the 32nd day of a prytany occurs several times (*IG* II[2] 683, 685, *Hesperia* 7. 1938 [125]), a date higher than 32 is not attested. Previous investigators of the calendar (Kirchner, Dinsmoor, Meritt) of this period had postulated prytanies which varied in length from 26 to 34 days.[6]

[6]Meritt's theories are multipartite and very confusing. I extract quotations for the topsy-turvy calendars he advocates in *TAPA* 95 (1964) 258: "The archon's tenure of office goes with the dates κατ' ἄρχοντα. But the prytanies with the untampered calendar ... The prytany calendar and the festival calendar began at different times. ... The κατὰ θεόν calendar, when dates are given in terms of it at all, is in harmony with the prytany calendar and presumably determines the beginning and ending of that year, not necessarily of the archon's year." He cites as an example a decree passed in Hekatombaion κατ' ἄρχοντα "would simply have been dated in the second

For ordinary years (354 ±1 days) of twelve phylai, we presented in *Calendars of Athens*, Chapter IV, pp. 78–88, tables showing the festival date, the prytany date, and the prytany date plus or minus the festival date. Several texts are dated on the 30th day of a prytany (p. 78), although no date higher than 30 occurs. Since in Table 1 showing triple dates (above p. 2–3) there are two cases where the κατὰ θεόν (= κατὰ σελήνην) dates show a deviation of one day from the prytany date,[7] we know that the lengths

---

prytany of that year. ... All κατὰ θεόν dates belong to the schematic years of the Metonic cycle." He postulates in other places that the festival calendar was one of alternating 30- and 29-day months. In the year preceding his example of a festival Hekatombaion = Prytany II, he posits a festival calendar of 13 months and kata theon and prytany calendars of 12 divisions. Meritt claims there was "never any period of dual archonships." This can only mean that this archon had two Prytany I's, one coinciding with Hekatombaion and one with Skirophorion of the same year. Thus, we have three calendars beginning with Prytany I 1 (the prytany, the kata theon, and the Metonic one omitting Day 64) and two with Hekatombaion (I) 1 (the festival and presumably the schematic one of alternating full and hollow months). Elsewhere, in *Hesperia* 38 (1969) 437, Meritt writes, "It began as a year of twelve months, which were divided into thirteen prytanies, and it ended as a year of thirteen months." This means that by Hekatombaion 1 or earlier, the officials decided that the ekklesia was to meet according to an intercalary year, but the festivals were to be held as for an ordinary year. In *AE* 1968 92–105, Meritt claims that the Metonic cycles governed his "lunar calendar of the prytanies," while in *Hesperia* 38 (1969) 438, it is the "lunar calendar of the archons" which was regulated by Meton. This kind of multiple theories is beyond parody.

[7]There are now two dates. At the time of *Calendars of Athens* (1947) there was one. We wrote (p. 78): "One might be tempted to conclude that there was exact coincidence between prytanies and the months of the civil <= prytany> calendar. The κατὰ θεόν dates in the second century, however, show a single deviation of one day from the prytany date. Though larger deviations are never attested, this is enough to force us to abolish the hypothesis of exact coincidence between the two calendars." Meritt/Traill, *Agora* XV.24, in defending the theory of a festival calendar with alternating 30- and 29-day months, err in writing, "Since the prytany calendar was always,

of the prytanies were not determined by the lengths of the months, which in the case of a lunar calendar was not determined until the end of the month. We noted that in some years the prytany date was ahead of the festival date, in others behind, but could formulate no rigid rule, and concluded:

> We feel that the problem of the intercalation and suppression of days in the civil <= festival> calendar, as determined from the fragmentary epigraphical evidence which usually furnishes only two or three inscriptions for any one year, is too elusive to permit the formulation of one rule for the sequence of 29- and 30-day prytanies throughout these two centuries.

Unlike the festival calendar which allowed extensive tampering in the form of embolimoi and exairesimoi days, the prytanies were of 29 or 30 days. Only the sequence is debatable. After much debate, Meritt in his last studies came to recognize this principle of only 29 or 30 days.

In an article, "Attic Civil and Sacred years," *Journal of Philology* 24 (1895/6) 76, T. Nicklin, citing the inscription now published as *IG* II^2 109, suggested that at the end of each prytany, the phyle to sit next was determined. In his "The Athenian Secretaries," *Cornell Studies in Classical Philology* 7 (1898) 20–27, W. S. Ferguson argued at length, after examining many texts, that this view was correct. The order of the phylai in prytany was not fixed by lot at the beginning of the year. Only in the ninth prytany in the period of ten phylai, for example, was the phyle to sit tenth known. On p. 27, Ferguson observed that in the year 408/7 the phylai in prytany were known for six prytanies and suggested that they were arranged in the reverse of official order. This observation was picked up by Meritt and Lewis and made the prime factor in their argument about the calendar of the period.

---

so far a we know, within its limits normal, it is of interest that the κατὰ θεόν dates of the festival calendar are always in accord with it."

See the commentary in *IG* I^3, bottom of p. 359. Meritt, while at Johns Hopkins, suggested the problem as a dissertation subject to a student who could find no evidence for officials in the reverse order or any other example in Attic inscriptions, and the dissertation was aborted. Once in several centuries, it is not surprising that six of the prytanies might be allotted in the reverse of official order or even in the regular order. The position of six proves nothing about the order of the remaining four.

In charting a list of about 150 equations for ordinary years after 307 B.C. (*Calendars of Athens* 78–88), we listed all prytany dates minus or plus the festival ones, but could find no pattern. No prytany date was higher than 30. At the time, the highest equation with an embolimos day was two. Moreover, on the assumption that what one archon did with his calendar another might, we did not have the advantage of Meritt's Table 4 above. There have been new texts and new readings since 1947, and other restorations could be offered exempli gratia. Some ordinary years have been taken to be intercalary and vice versa.[8] In the light of the Athenian passion for allotment, and since a modicum of pay (5 obols) was involved,[9] it may be that just as the lot determining the phyle in prytany throughout the course of the year, so with the choice of 30 or 29 days in some fashion that escapes

[8]In the latest archon table for part of the third century (M. J. Osborne, *ZPE* 58 [1985] 275–295), the author disregards the problem of the sequence of ordinary and intercalary years, and concludes that the secretary cycle did not operate in the absence of democracy, although adopting the theory of Meton's cycle in *AJA* 103 (1999) 710–711. The long-awaited promise (Osborne says since 1992) by J. Morgan for a new archon table will be valueless unless the author establishes the character of years on a sounder basis than exhibited by Meritt in *Historia* 26 (1977) 161–171 or in Woodhead's *Agora* XVI.

[9]See W. T. Loomis, *Wages, Welfare Costs and Inflation in Classical Athens* (1998) 26, for the pay of bouleutai. Those serving as prytaneis received an additional obol.

us. The subject might be re-opened with a new study of calendar equations on the assumption that what one archon did, another might.

Calendar Equations. Turning to the restoration of calendar equations in prescripts of decree, which now must number over 300, we offered in 1947 texts and charts with restorations which would be in accord with the Aristotelian pattern, focusing on the period of Aristotle (before 307/6) and intercalary years. In addition to irregularities proved by Tables 1, 2, and 3 in Chapter 1 above, we sometimes assumed mason's irregularities for which we could cite parallels. We worked under the assumption that what one scribe or stonecutter could do, another might. As hypothetical examples, one might assume the spellings πρυτανέα, ὀγδοίης, νέια, if other factors were similar.[10] The thrust of Meritt's *Athenian Year*, in addition to trying to reject the theory of a lunar calendar, was to disprove the Aristotelian pattern, and that solely by the restorations of calendar equations without addressing other arguments. Meritt's approach was a double standard. If we, in support of the pattern, restored an irregularity, such as a vacat, where Meritt's competing text required none, he claimed his interpretation of the year was superior; whereas for another text, he restored the same type of irregularity in claiming the rejection of the rule. In *Athenian Year*, he assumed seven errors in the names and days of months and prytanies, whereas we assumed five, but in some cases different ones, which he then discredited. He makes a complete u-turn and reverses himself on positions he claimed earlier he had proved, stating that there is "a lack of evidence", a phrase repeatedly used.

I offer in Table 6 a catalogue of the irregularities which Meritt assumed in his *Athenian Year* (alone) in restorations of calendar

[10]See L. Threatte, *The Grammar of Attic Inscriptions* 1 (1980) 316, 334, and 152.

equations. In *Ancient Athenian Calendars on Stone* 375 ff., I quoted the very revealing justifications which Meritt offered in defense of many of his restorations.

TABLE 6

IRREGULARITIES IN CALENDAR EQUATIONS FROM MERITT'S *ATHENIAN YEAR*

	Page
Omission of word	52, 55 (and 126)
Omission of 2 words	97
Haplography (omission of letter)	70
Uninscribed letter-space	86, 96, 113, 123, 164 (*bis*)
2 uninscribed letter-spaces	126, 127, 164
2 letters in one letter-space	55, 70, 84, 105, 126, 127
Error in number of prytany	84-85
Error in day of prytany	112, 161
Error in name of month	111, 121
Error in day of month	93, 158
Calendar phrase misplaced	100
Use of unattested word or phrase not attested in the formula	57 (*bis*), 70
Whole stoichos uninscribed except in line having calendar equation	119, 164

In one inscription, *IG* II^2 1672 (329/8), Loomis, *Hesperia* 64 (1995) 133, following Kirchner, has reported 43 inscriber's errors including omissions and "letters, numerals and words inscribed by mistake." In Table 5 of *Stone* p. 310, I listed 20 textual anomalies which occur on the stone in one beautifully cut inscription.

Scholars who focus only on the restorations of equations averaging about two per year find it easy to weigh the competing versions and pronounce that festival plus prytany calendar correct which can be restored with no irregularities, or the least number, without surveying all the ingredients of a complex problem. This is the practice of Woodhead throughout *Agora* XVI. A survey of the entire picture is required. We are not dealing with a perfect medium.[11]

The determination of the character of most years depends almost invariably on restorations often with debatable readings or disputed letters. This includes the sequence of prytanies, the subject of this chapter, as determined by their correlation with the festival calendar. In his debate on the subject, Meritt formulated rules about the position of irregularities, just as we have seen above about embolimoi days.[12] One such rule was put forth about the position of vacats in restored equations. Meritt claimed repeatedly that a stonecutter would never leave space between the day of the month and the day of the prytany. In *Calendars of Athens* we had restored a vacat in offering an equation which would allow a sequence of prytanies according to the Aristotelian pattern. Meritt's "rule" has found currency with those not accepting the

[11]R. P. Austin, *Stoichedon Style* (Oxford 1938) 47, wrote of *IG* I² 359 (= *IG* I³ 455), "The number of letter-spaces left vacant at the ends of the lines in this inscription is wrongly represented in *IG*. i². The correct account is as follows: normal line = 20 letters; vacant spaces—one at the end of lines 7, 9, 10, 11, 15, 20; two, line 13; three, line 8; an extra letter at the end of lines 12, 14, 19, possibly 21; in line 7 the letters λεμμ on the left are extra-stoichedon." In Chapter VIII on the decline of the stoichedon style in the third century, Austin attributes this to the economic factor (pp. 104–105); a stoichedon text is more costly in material, time, and skill than a non-stoi-chedon one. Also, tapered stelai deterred the engraver from using stoichedon texts. W. T. Loomis, *Wages* (1997) 121–165, has a valuable chapter on the payment formulas in Attic inscriptions.

[12]Dow, *HSCP* 67 (1963) 61–67, discusses the many instances of Meritt's "irrational" gaps, a subject on which there will never be agreement.

Aristotelian pattern. M. J. Osborne, *Naturalization in Athens* 2 (1982) 103, writes:

> PRITCHETT and NEUGEBAUER have taken the year to be intercalary and have proposed to restore the date as ἕ[νηι καὶ νέαι v δευτέρ]αι καὶ εἰκοστεῖ τῆς πρυτανεία[ς] (i.e.*Posideon* (VI) 30 = Prytany V 22 = Day 178). Against this is the hypothetical nature of the vacant space assumed between the date by month and the date by prytany. Such a *vacat* is not, of course, impossible, given the vagaries of the usage of vacant spaces in the prescripts of decrees, but it is to be noted that punctuation by means of vacant spaces in this document is minimal otherwise and further that this would be an extraordinary place to make a pause. More generally, it is questionable whether it is sound procedure to hypothesize such an irregularity in the *stoichedon* pattern, if a restoration can be made without such recourse.

The inscription is *IG* II^2 448 with two decrees, 323/2 and 318/7. The text in question is the first decree, lacking the all-important right margin. I offer a history of the text of line 3, following the word Ποσιδεῶνο[ς]:

Koehler, *IG* II.5.231b	=	ἕ[κτῃ ἐπὶ δέκα, δευτέρα] with Lolling's majuscule text showing epsilon followed by two uprights.
Kirchner, *IG* II^2 448A	=	ἕ[κτηι ἐπὶ δέκα, δευτέρ] Ordo στοιχηδόν saepius turbatus est: V.10.52 = 40 litt., 59. 67 = 41. || In v. 56 in. litt. M H in uno spatio incisae sunt, item 58 med. A I (αἶς), item 86 med. Ω N.
Dinsmoor, *Archons of Athens*	=	ἕκτ[ηι μετ' εἰκάδας, δευτέρ]α "giving only 40 letters in line 3 (as in lines 10, 52)"
Calendars of Athens	=	ἕ[νηι καὶ νέαι v δευτέρ] or [νείαι]

Meritt, *Year* 109	=	ἕκ[τηι ἐπὶ δέκα, δευτέρ] "The kappa in the word ἕκ[τηι, about which there has been doubt is entirely clear—all three strokes."
Klaffenbach, *Gnomon* 21 (1949) 135	=	"Ich habe den Abklatsch von IG II/III^2 448 geprüft. Er ist sehr schwer lesbar, aber wenn ich mich auch nicht dafür verbürgen kann, so muss ich doch sagen, dass ich hinter E in der Tat ein N zu lesen glaube, aber weiter dann auch nichts. Jedenfalls kann ich Kirchners Lesung ἕκ[τηι nicht verifizieren."
Meritt, *TAPA* 95 (1964) 227	=	ἕκ[τει ἐπὶ δέκα, δευτέρ] "From my own examination of the stone following the kappa, I could find no stroke on the stone, yet a stroke as of epsilon, had epsilon been there, should have appeared. ... Pritchett, in order to make an intercalary year with ἕ[νει, must leave an uninscribed space in *IG* II^2 448 between the date by month and the date by prytany . ... I shall discuss elsewhere how rare an uninscribed space at this point in the preamble of a decree must be."
Hansen, *GRBS* 23 (1982) 346 no. 71	=	ἐ[νάτει ἐπὶ δέκα δευτέρ] with an extra letter, citing lines 56 and 57.
Schwenk, *Athens in the Age of Alexander* 404	=	"Pritchett–Neugebauer's restorations, however, depend on a new reading of st. 25 and st. 26 in line 3 of #83. Their restoration of a nu and eta is unacceptable when kappa and tau can be read clearly (see commentary, #83). Thus the restoration in II^2 448 must be retained (lines 3/4): ἕκτ[ηι ἐπὶ δέκα, δευτέρ]/αι καὶ εἰκοστεῖ τῆς πρυτανεί[c·---]."

Her commentary gives no description of the "clear" letters, but reports of readings read by no other epigraphist are not uncommon in this book.

As reported in *Stone* 285–186, in 1951, I was given a ladder to inspect the stone stored in the National Museum. I reported, "After the first epsilon is an upright stroke from the top of which projects a slanting stroke" as of nu. In 1962, I returned with Mitsos and Stroud and reported, "There remains no doubt about the upright hasta in the left part of the letter-space", but other traces were elusive. All that we can be sure of is the epsilon.

In the above prescript, there are 17 letter-spaces to the right of the epsilon. Writing in 1947, I made no effort to study the position of vacats and placed one between the two dates. Dinsmoor was wiser; he simply gave 40 letters to the line. The attack on the position of the vacat is sophistry. It may be placed at the end of the line for which there are many parallels.[13] On the same stone in the second decree, Meritt in *Year* 126 restored the date as: ἕ[κτει <μετ'> εἰκάδας]. On the same page and in the same year, he restored line 2 of a text as ἧς Κεκροπίδ[ος ἕκτης vv πρυταν]. Surely, there is closer unity between an adjective and a noun than between two dates of the calendar equation. He wrote, "The corruption of the stoichedon order in line 2 was caused, apparently, by the cutting of κροπίδος instead of Κεκροπίδος." But his photograph shows no rasura, and, as an experienced editor, he prints no siglum for such in his text of the line.

[13]Meritt's double-standard is illustrated by the fact that after Dow had criticized his restoration of [τριτηςv], he wrote (*Hesperia* 32 [1963] 432), "One need not insist that the uninscribed space followed τρίτης immediately; it may have come before or in the word or elsewhere within the line. The important point is not which stoichos was left blank but that somewhere in the line there may have been finally only six letters in a space of seven stoichoi. ... This could have resulted from the erasure of a wrong numeral ... It is best not to rationalize too closely about what must have been, *ex hypothesi*, an irrational irregularity; the range of conjecture is too wide."

This one prescript of *IG* II² 448A illustrates the lack of agreement on equations which must be resolved before we can establish rules about the length of prytanies from calendar equations. One can set up a straw man and offer reasons for the validity of any single restoration which seriatim sounds convincing, just as with embolimoi and exairesimoi days,—rules about vacats, errors of stonecutters and secretaries, extra letters, limiting observations to selected lines, physical criteria, etc., none of which cannot be paralleled from other stones.[14] The debates over fragmentary letters will continue.[15] Using the irregularities in Table 4 in Chapter 1 and in *TAPA* 95 (1964) 253–254 and following Meritt's computations that a month might be as long as 42 days and as short as 14, there are virtually no determinations in calendar equations for the length of prytanies in any given year which may not be charted to conform to the Aristotelian pattern. Obviously, any equation giving the day in the year depends on the correlation of the two calendars, and we believe the festival calendar was lunar. The salient fact remains that in the hundreds of texts where a prytany date is given no number has turned up which exceeds that of the Aristotelian pattern for any given pry-

[14]Meritt wrote in *Athenian Year* 56, "As every epigrapher knows, the stoichedon order is never a formal bar to a good restoration, even if a violation has to be assumed in an otherwise perfect text."

[15]By usual standards dots under letters are to be used when the traces do not exclude other letters. With Meritt and others, they have frequently come to imply letters partially preserved where no other reading is possible. Thus in *IG* II² 448B, where Kirchner has [ἕκτει ἐπὶ δέκα], which we followed in *Calendars of Athens* 64, Meritt (*Hesperia* 43 [1974] 463) reads ἕ[ν]η̣[ι καὶ νέαι] which he earlier called "epigraphically irreproachable" (*Year* 56) and in *TAPA* 95 (1964) 254 claimed that there was an upright in the right part of the letterspace, but he had not seen this when he inspected the stone (*Year* 57 n. 32). The upright was created out of thin air. Osborne (*Naturalization* I.103 and II.105) expanded the eta to ἕν̣η̣[ι without explanation of the two dots. The reader is lulled into thinking that the word with the dotted letter is the only possible one.

tany. Meritt (*Agora* XV [1974] 24) seems to make something of a u-turn in writing, "The prytany calendar was always, so far as we know, within its limits normal". The word 'normal' is defined in Webster as "according to, or not deviating from, an established norm, rule, or principle".[16]

In his 1997 volume, Woodhead (*Agora* XVI p. 128) finds one year (327/6) where he affirms positively that the Aristotelian pattern does not apply: "The necessary restorations in *IG* II^2, 357 show that Aristotle's 'rule' cannot in any case be maintained for this year, unless an error on the part of the stonecutter be postulated." The proviso is not without interest. In table 14 on p. 374 of my *Ancient Athenian Calendars on Stone* (1963), repeated in Table 6 above, I collected irregularities assumed by Meritt in his *Athenian Year* for his restorations of calendar equations in fragments of texts almost all of which were labeled stoichedon. These irregularities were discussed in detail in the following pages. Woodhead's position throughout his book is to accept Meritt's assumed irregularities in order to reconstruct a calendar violating the Aristotelian pattern, but any license for the assumption of error, even ones well-attested on stones, by a scholar supporting the pattern is intolerable. Meritt, for example, assumes 42 days for one month and 14 for the next without Woodhead raising an eyebrow. On p. 378, I noted that I had offered in *Calendars of Athens* an equation for *IG* II^2 357 in which I suggested that in the restored text the word ἦι was omitted, citing *IG* II^2 242, *Hesperia* 7 (1938) p. 292, and parallels of omission in Meritt's table. The stone in question is a small fragment with parts of seven lines preserving at its maximum width eight letter-spaces in one line.[17]

[16]Cf. Meritt, "Normal Lengths of Prytany in the Athenian Year," *GRBS* 17 (1976) 147–152.

[17]Only the day of the month is preserved. Meritt and Woodhead fail to tell the reader that we offered other restorations for both intercalary and ordinary years. The smaller the fragment, the more the restoration, the more

Kirchner found that two lines (2, 4) gave restorations for a stoichedon 29 text with line 6 of 28 letter-spaces. The name of the month was clearly omitted by the mason. In *GRBS* 23 (1982) 344, M. H. Hansen commented on the text:

> The number of the prytany has to be restored and the name of the month is omitted; various restorations and equations have been proposed. What concerns us here is the day of the prytany, ἕνηι κ[αὶ νέαι . . . . . . τῆς πρυταν]είας. Six spaces give us a choice between πρώτηι, τρίτηι, ὀγδόηι, and ἐνάτηι. If we assume a slight violation of the *stoichedon* order, we can extend the possibilities to cover πέμπτηι, ἑβδόμηι, and δεκάτηι. The only inference to be made is that the meeting was held early in the prytany.

Similarly, in line 3, where ἕκτης is restored for the number of the prytany, one may say that a violation of the order of lines 2 and 4 would give us the possibility of τρίτης, ὀγδόης, and ἐνάτης. All that is certain is that an early day in a prytany was equated with the last day of a month.[18] On p. 54 of *Calendars of Athens*, I of-

positive Meritt and Woodhead become. In *Historia* 21 (1977) 169, Meritt marks the character of the year with an asterisk! However accurate his archon table, many asterisks must be removed. Meritt gives only his own bibliography.

[18]S. Dow, in his article on "Calendric Theory vs. Epigraphical Fact" in *HSCP* 67 (1963), dealt with another text (*IG* II^2 358) from this same year (pp. 57–60), which he humorously shows might belong in another year, since the archon and secretary are complete restorations. For the date, see also *SEG* 26.87. The difficulty with Woodhead is that he does not understand the meaning of square brackets. I discussed the difficulties of yet another text from this year published by Meritt, *Hesperia* 3 (1934) 3–4, in *Ancient Athenian Calendars* 273 ff. So far no restoration for a lacuna of 19 letter-spaces between the name of the month and the day of the month has found general acceptance. I proposed an attested phrase with the word ἡμερολεγδόν in the count with μετ' εἰκάδας which would fill the lacuna. Meritt, typically, countered by manufacturing a 'rule' that only the scribe of 307/6, where the word is attested, could use the word, but does not tell us why. A better 'rule' is that the word was used only to indicate the count with μετ' εἰκάδας.

fered a reconstruction of the calendar for an intercalary year in which the Aristotelian arrangement of prytanies may be followed by assuming that the festival calendar was retarded by three days in the middle of the year, and that without assuming any irregularity on the part of the mason.

Dow, *HSCP* 67 (1963) 67–75, analyzed five examples of restorations for one prescript of *Hesperia* 13 (1944) 236 (320/19) offered by Meritt. This fragment has neither margin; where the lines begin is a critical issue. Meritt, as editor of a periodical which published only articles approved by him, immediately replied in *Hesperia* 32 (1963) 425–432 offering a seventh. His calendar equations for the year result in an unparalleled sequence of the prytanies and a festival calendar with months of 42 and 17 days. In *Ancient Athenian Calendars on Stone* 377, I concluded that all that is acceptable is what is on the stone, a day between 2 and 10 of an unknown month was equated with a date ending in 4 of an unknown prytany.

Prytany Calendar: Fifth Century. When Neugebauer and I submitted our manuscript of *The Calendars of Athens* to the Publications Committee of the American School advocating the theory of the Aristotelian pattern in *Ath. Pol.* 43.2 for the prytany calendar in the fourth and following centuries,[19] we were told that it should be tested for the fifth century in order to be valid. As a result, we added Chapter V to our manuscript, reaching a conclusion in

[19]Meritt and his followers (e.g., P. J. Rhodes, *Commentary on the Athenian Athenaion Politeia* [1981] 519: "Meritt believes that the lengths of prytanies did not vary more than necessary, but A.P.'s placing of the longer prytanies at the beginning of the year is merely an illustration") do not explain why a pattern of irregularity would be applied or produce one "necessary" example outside of the manipulation of calendar equations. Even if we did not have Aristotle, one might expect a calendar determined in advance not only for the year but for each prytany, and without embolimoi days, to have equitable parts.

support of an extension of the pattern. Calendar equations were not used in the prescript of decrees; there are few dates in terms of the festival, or lunar, calendar before 407 B.C.; and scholars differed on the sequence of ordinary and intercalary years with no scholar finding a fixed order of such years. In other words, the relation of Hekatombaion 1 of any year to the Julian calendar is a matter of conjecture. Whereas the festival calendar had been used as the basis of all reconstructions, our theory turned to the prytany calendar.[20]

Few now contest the theory that the length of prytanies was determined in advance. Such phrases as "seven days were left in the prytany" (λοιπαὶ ἐ͂σαν ἑπτὰ ἑμέραι τε͂ι πρυτανείαι) occur (*IG* I³ 369.6). The last day of the prytany was τελευταία (*IG* I³ 364.22). See the numerous prytany dates in I³ 365. The highest date preserved in any given prytany provides the minimum number of days for a prytany. In I³ 375, we have 36 days for prytany 8 (line 29), prytany 9 (33, 34) and 10 (39). As observed above, in the Erechtheion accounts, we have 37 days in prytany 6 and 36 in prytany 8. A priori, one would expect an equitable division. After testing cases where the length of a prytany could be determined from texts not involving restorations, we advanced the following table for the fifth century corresponding to the later Aristotelian arrangement (*Calendars of Athens* 97):

Number of Prytany	Number of Days in Prytany	Number of First Day of Prytany	Number of Last Day of Prytany
I	37	1	37
II	37	38	74
III	37	75	111
IV	37	112	148
V	37	149	185
VI	37	185	222

[20]See "Calendars of Athens Again," *BCH* 81 (1957) 269 ff.

Number of Prytany	Number of Days in Prytany	Number of First Day of Prytany	Number of Last Day of Prytany
VII	36	223	258
VIII	36	259	294
IX	36	295	330
X	36	331	366

For the epoch year of the introduction of the prytany calendar, Cavaignac, *REA* 31 (1929) 213, favored the year 502/1, a year of the Great Panathenaia, because of the passage in *Ath. Pol.* 22.2 stating that an oath of office was instituted for the Boule of Five Hundred. As Rhodes points out, the working out of the details of Kleisthenes' reforms required several years. Pollux (8.110) places the institution of ten phylai (δέκα ἐγένοντο) in the archonship of Alkmeon (502/1: Rhodes). As will be explained later, such institution might take place any time within the archonship. There were other fixed solar phenomena of 365 and a fraction days than the summer solstice (the first day of Cancer). The Geminos parapegma, as well as others, divided the solar year according to zodiac signs. The ten phylai must have had some sort of equitable calendar. The ten generals were chosen for ten phylai at Marathon, and Capps (*Hesperia* 12 [1943] 11) urged that tribal competitions in the Dionysia began in 502/1.

Early fifth-century decrees (c. 460) name a prytanizing phyle in the prescripts (*IG* I³ 7 ff.). At the time of the institution of this ten-phylai boule, there presumably was some regulation about the length of each prytany. Meritt, *Athenian Year* 203 n. 4, advocated the time of Kleisthenes for the institution of what he termed 'the seasonal conciliar year', objecting, however, to a rigid prytany year of 366 days because its epoch date would be 'nearer to the vernal equinox than to the summer solstice', otherwise giving no clue to its early nature except to generalize that 'the Council began its term about midsummer'. Rhodes (*CAH*² [1992] 79) attributes the institution of the prytaneis to Ephialtes. Prescripts with the name of the phyle in prytany occur in *IG* I^3 7, 9, 10, 11, 17, 18, 21, 23, 27, 28, 30, 31, 34. The dates of many inscriptions are contested, but are dated in the corpus, often on the unreliable basis of script, to 460 or later. From the time given by Pollux to Ephialtes, we have no epigraphical evidence. A secretary was in place about 500 B.C. (I^3 5). I favor the 502 date.

In a very strange article which has crept into a periodical founded by Gildersleeve, Francis M. Dunn, "The Council's Solar Calendar," *AJP* 120 (1999) 369–380, regards the calendar of 4 × 366 days in the Logistai inscription (to be discussed below) as a poor approximation of the solar year, what he calls "a huge degree of error", oddly, without reference to Meritt and McGregor, attributing the calendar to me with calculations that I have not offered (p. 371):

> The problem with Pritchett's theory is that it is highly implausible. Why would the council adopt a calendar that was a poor approximation to the solar year? When the council took the radical measure of adopting a calendar that began and ended at different times from the archon's calendar, it would surely have taken the trouble to determine the length of the solar year more precisely. One solution is to push back the introduction of the independent year to the time of Cleisthenes, when estimates of the solar year would have been much less precise than in the late fifth century; this seems to be Pritchett's preference, since he argues against a late introduction of the independent year (Pritchett and Neugebauer 1947, 105–6) <*sic*>. The difficulty with this solution is the huge degree of error it involves. In four years, the calendar would fall three days behind the solar year, so that by 468/7 the council's calendar would be a month behind the sun, by 429/8 it would be two months behind, and by the end of the century, when it was finally abandoned, the council's calendar would be more than two and one-half months behind the sun. Furthermore, this discrepancy remained uncorrected, since the quadrennium 426/5 to 423/2 still av-

> eraged 366 days per year. A second solution is no better. We might assume that the council's independent calendar was only introduced in the late fifth century, and that in the quadrennium 426/5 to 423/2 it had not yet diverged very far from the tropical year. But this would mean that the poor approximation of 366 days was adopted at a time when it was known to be wrong, after Meton had already published his much better estimate of 365 $^{5}/_{19}$ days. Finally, we might attempt to avoid these difficulties by arguing that the independent calendar was intended to be purely schematic, with an arbitrary period of 366 days that only by chance approximated the tropical year. Yet this is not tenable: all Greek calendars were designed to follow the cycles of the moon, sun, and stars, and there is no evidence in any period for a Greek calendar permanently out of phase with both the moon and the sun. The most plausible solution is to follow Meritt in regarding the council's independent calendar as an approximation of the solar or tropical year.

I have marked *sic* where Dunn misinterprets me, as he does in all three articles he has published. It is regrettable editors do not require quotation marks. In the records for the Athena Promachos statue (c. 450), Meritt (*Hesperia* 5 [1936] 368–377) attempted to extract salaries, per diem by the prytany calendar, from the text now published as *IG* I^3 435 lines 81–82, and claimed (p. 376) that they proved that pay was for 380 days, which he regarded as an approximation to a prytany year of 384 days. Then, he made a leap into space and posited that this prytany year of 384 days proved that the two calendars were coterminous. Both were of 384 days, as for an intercalary year. "We have to deal in the present instance with an intercalary civil year of thirteen months coterminous with the year of the ten prytanies."[21] In *Calendars of Athens* 105–106, we denied that this interpretation of the figure was evidence for a prytany calendar of 380 (or 384) days, and, since pay

[21]Meritt's position was that c. 450 B.C. the prytany calendar was tied to the festival one.

was made by the prytany, suggested that the sum was the product of 10 × 1178. This is now academic, since W. T. Loomis, *Wages, Welfare Costs and Inflation in Classical Athens* (1998) 97, has devoted a page to showing that Meritt's computations were all incorrect. In *Calendars of Athens*, we limited ourselves to evidence about the pattern of prytanies. The essential fact remains that in those cases where we in 1947 found pay by the prytany, they all conformed to the Aristotelian pattern, and this now holds for all the examples listed in Loomis.

In the first part of the quotation, Dunn objects that the Athenians could not have had a calendar of 366 days because it would be out-of-step with a solar calendar. He assures us that it was "highly implausible" that the Athenians could have adopted a prytany year of 366 days because it strays from Meton's solstice year of 365 5/19 days; they knew it would be wrong. One might anticipate that the author was going to offer a solution that the prytany calendar was not inaccurate to 14/19 of a day. But then we are told in the same quotation that Meritt was right in regarding the prytany year as an approximation. Meritt, in the course of his research, had posited years of 362, 365, 368, figures which are more inaccurate than 14/19 of a day.

Francis M. Dunn proposes several solutions. First, he writes (371–372), "To keep this calendar aligned with the sun would frequently require the suppression of one day ... This omission, I suggest below, would not occur on a rigid schedule, but would be controlled by observation of the summer solstice." Next, Dunn says that the number of days loans were outstanding in the Logistai inscription does not represent the actual number in the Athenian prytany calendar; the abacus-operator "rounded off": "The simplest way to calculate interest would be to reckon each year as 366 days. Such rounding-off would not be surprising ... It would be reasonable and unsurprising to use 366 as a round estimate for solar years that sometimes had 366 days and sometimes 365."

Finally, he says (375–378) that this calendar of 366 days is to be attributed to Meton, "who established a tropical year of 365 5/19 days." Dunn does not mention that the earlier octaeteris in Geminos 8.27–33 was constructed on a solar year of 364 1/4 days.

On p. 373, Dunn writes:

> Such rounding-off would not be surprising. As Meritt noted, the accounts for the statue of Athena Promachos seem to use ten 38-day prytanies as a round estimate for a 384-day year in calculating payments (*IG* I³ 435 81–82 [ca. 450]).

On the following page, he writes:

> Meritt showed that in the accounts for the statue of Athena Promachos (*IG* I³ 435 81–82 [ca. 450]) the annual salary for the supervisors was calculated for a period of 380 days (380 days times 31 obols per day = 1,963 dr. 2 ob.). It is therefore likely that at this date, circa 450, the round figure 380 was used to approximate the 384-day intercalary lunar year (380 = ten 38-day prytanies) and not the 366-day solar year.

In three articles, Dunn shows no first-hand acquaintance with epigraphical documents. Everything is ex cathedra pronouncements about second-hand appraisals. Not only was Meritt completely wrong, but in *IG* I³ 435, we are repeatedly told μισθοὶ κατ' ἐμέραν, μισθοὶ κατὰ πρυτανείαν, μισθοὶ ἀποπάχς (lines 18/19, 28/29, 76/77, 111/112). Only the prytany calendar was used. Any connection with the festival calendar is an ignis fatuus.

Dunn cites no example of "rounding off" with the prytany calendar. The Logistai inscription is an unacceptable candidate. The interest was 1/4 obol per diem in 1250 drachmai, and was computed through the quadrennium. The terminology "so many days were left in the prytany" (λοιπαὶ ἔ͂σαν ...) bespeaks an exact count of days. Dunn avoids anything specific to support his generalities. On p. 372 n. 7, he writes, "My argument does not require that the longer prytanies come first." He deduces no example where the Aristotelian pattern is proved not to apply. In *ZPE* 123 (1998) 225, he again generalizes without documentation, "A date in the council's calendar may vary by several days."[22] Much has been made of "Who Killed Homer?", but a more pressing question is "Who Killed Training in Classical Research?" at some of our universities.

Starting at least with Dinsmoor, the shadowy Meton has been made the principal actor in Athenian calendric studies, which include the prytany calendar. Dinsmoor theorized that in 432 B.C. the Athenians made Prytany I 1 = Skirophorion 13 = solstice. Meritt (e.g. *AFD* 152 ff.), although concurring that Meton was responsible for the Athenians instituting a solar prytany calendar, vigorously attacked this equation on the basis of an intricately reconstructed calendar from a few widespread dates. He calculated that in 432, Prytany I.1 = Skirophorion 18. By Meritt's calculations, the prytany solar year never began with the solstice, but was not far removed. Dunn, whose work is flawed by a mismatch between his vaulting theoretical ambitions and his faulty grasp of the details of the astronomical and epigraphical material, assures us that the Athenians had a solar observation prytany calendar, determined by the solstice, but, ever frowning on details, gives no clue as to the epoch date in terms of the festival or Julian calendars, except to reject Dinsmoor on p. 377.

There are three related problems, epoch date, solstice, and Meton, which are treated briefly.

Epoch Date of Prytany Calendar. Toomer, in *Dictionary of Scientific Biography* IX.338, wrote, "One can begin a cycle at an

[22]The prytany calendar is not to be confused with the festival one. Heath, *Aristarchos of Samos* 285–286, along with Ginzel, showed that a month in the festival calendar was often referred in popular parlance and computed as of thirty days. Other examples could be added. See Chap. III n. 9 above, and A. Mommsen, *Chronologie* 49–53.

arbitrary point." E. J. Bickerman, *Chronology of the Ancient World* (1968) 41–42, writes:

> Modern scholars inferred, without any warrant, that the Egyptian annus vagus must have started on the day when the heliac rising of Sirius fell on 1 Thot. From Fréret (1758) on, they hesitated between 1322 and 2782 as the starting years of the Egyptian calendar (Ideler I, 126). Ed. Meyer went back to 4241. But the dispute is futile. A calendar is a tool which cannot be justified by either logic or astronomy. The Egyptian calendar took account of the Nile and not of Sirius (O. Neugebauer, *Acta Orientalia* 1938, 169; *JNES* 1942, 396). Furthermore, there is no inherent necessity to start a new calendar on its first day. England changed from the Julian calendar and the year beginning on 25 March, to the Gregorian style and the adoption of 1 January as the New Year, on 2 September 1752.[23]

As noted above, Pollux is our only specific source for an epoch date. I favor a date of c. 502 B.C.[24] After about 460 B.C. the name of the prytanizing phyle occurs in our earliest prescripts. Rhodes favors Ephialtes. Apparently, a decision was made that the calendar of ten parts (phylai) was to equal a solar year to a fraction of a day. The citizen body may have hit upon one of many astronomical phenomena. The early Greeks attached significance to the risings and settings of fixed stars and star-groups (Pleiades, Hyades, Orion, Sirius, Arcturus, etc.). Plato, Aristotle, Theophrastos all use risings of stars and solstices to specify times of year. See, for example, M. L. West, *Hesiod Works and Days* 376–381. In *BCH* 85 (1961) 32–37, Van der Waerden and I published

[23]In the church calendar, Gregory directed the day following the feast of St. Francis, that is to say the 5th of October, to be reckoned the 15th of that month. See the article on the calendar in the *Encyclopedia Britannica*, 11th edition.

[24]As noted in *Greek Archives, Cults, and Topography* (1996) 25, T. L. Shear, Jr. dates the Old Bouleuterion about 500 B.C.

Euktemon's astronomical calendar listing the dates of rising and setting of stars, computing Julian dates.[25] Hesiod and Demokritos identified the beginning of spring with the evening rising of Arcturus. In treating the solar year of Thucydides in *Thucydides' Pentekontaetia* (1995) 197, I cited the seasonal calendars of various authors, including the Hippocratic *De regimine* of the fifth century; Arkturus and Pleiades.[26] Much material is given in M. Nilsson, *Primitive Time-reckoning* (Lund 1920). In adopting a calendar approximating a solar year within a fraction of a day, if our conjecture is correct, the Athenians, like Gregory, may have begun on a day after a festival. Once it was set in motion, there was no reason to modify it. It was not tied to any astronomical event, and its epoch day was not commemorated. Later, in the period of our many calendar equations, it was tied to Hekatombaion 1.

A. J. Toynbee, *A Study of History* 7 (1954) has much to say about Time and Calendars. On p. 298, he writes,

> The dates chosen as the starting-points for new eras had in some cases been those of events of which the authenticity and the time of occurrence had been established beyond dispute. Classical examples were the eras starting from the Fascist occupation of Rome on the 28th October, 1922; from the establishment of the First French Republic on the 22nd September 1792; from the Prophet Muhammad's *Hijrah* from Mecca to Medina on the 15th July, A.D. 622; from the assumption of a formal claim to oecumenical authority in the Indic World by the Gupta Dynasty on the 26th February, A.D. 320; from the definitive establishment of the Seleucid Empire's Hasmonaean successor-state in Judaea in 142 B.C.; etc.

[25]See Rehm, *RE* Suppl. 7 (1940) 175–198 s.v. Episemasiai. For the Julian dates of some of the annual fixed stars, see F. K. Ginzel, *Handbuch* II.520 ff.; Bickerman, *CAW* 1st edition 143 ff.

[26]See Böker, *RE* Suppl. 9 (1962) s.v. Wetterzeichen 1610–1692.

On the other hand, some eras are reckoned from events which are disputable. There is no evidence that Jesus was born in the first year of the Christian era. The traditional date was instituted in A.D. 525 at Rome by the Abbot Dionysius Exiguus at the instance of the Pope. The original provision in the United States for a four months' interval between the election of a President's electors and the President-Elect's assumption of office was made with the practical purpose of ensuring that he had time to convey himself, family, and possessions by horse-transport from his native state, even if Georgia or New Hampshire. The invention of transportation by railroad stultified this practical purpose, but it was not until 6 February 1933 that the interval was shortened by the Twentieth Amendment. One can think of many epoch-making events which were never taken as starting-points of new eras, such as the landfall of the Pilgrim Fathers on 21 December 1620.

Solstice. Dunn makes one departure from Meritt, claiming that the prytany calendar was "controlled by observation of the summer solstice" (p. 372). Whether he means that the prytany calendar ran from solstice to solstice or that observation gave an 'advance notice' is not made clear. Difficulties with observation of the solstice make this theory highly improbable, even if we did not have calendric considerations.

The summer solstice is the moment of the sun's greatest northward displacement from the equator, and thus may occur in the night as well as in the day. James Evans, *The History and Practice of Ancient Astronomy* (1998) 205, writes:

> There are severe limitations on the accuracy with which a solstice may be determined. The Sun's motion in declination is so slow around the solstices that the moment of greatest declination is very uncertain. As the table of obliquity (table 2.3) reveals, after the summer solstice, the Sun must run 6° or 7° along the ecliptic for its declination to decrease by only 10′. So, for a week on either

> side of the solstice, the length of the noon shadow scarcely changes. The gnomon itself produces an additional uncertainty: the edges of shadows are fuzzy and indistinct, so a precise location of the shadow's tip is impossible. This fuzziness arises from the fact that the Sun is not a geometric point, but has an appreciable (1/2°) angular diameter.

A. C. Bowen and B. R. Goldstein, "Meton of Athens and Astronomy," in *A Scientific Humanist: Studies in Memory of Abraham Sachs* 48, ask, "How could Meton know when the solstices were to occur?" and reply that he obtained the knowledge, not by observations, but from the Uruk scheme of the Babylonians.

As to Athenian observers determining the time from solstice to solstice, E. J. Bickerman, Chronology of the Ancient World[2] (Ithaca 1980) 30, writes:

> Censorinus, ... explains the disarray of pre-Julian lunisolar calendars by the uncertainty concerning the actual duration of the solar year. In fact Hipparchus (c. 125 BC) still had to oppose the opinion of those astronomers who believed that the length of time in which the sun passes from a solstice to the same solstice again is exactly 365 $\frac{1}{4}$ days (Ptol. *Almag.* III, 3). Hipparchus himself was able to give the almost exact value of the length of the year (365 $\frac{1}{4}$ – $\frac{1}{300}$ of a whole day) which is less than 7 minutes in excess over the true mean year (*cf.* T. Heath, *Aristarchus of Samos* (1913), 297). Yet he acknowledges the possibility of error in the observations, which according to him could amount up to 3/4 day for the time of a solstice and up to 6 hours for the time of an equinox. Thus too Ptolemaeus, who quotes Hipparchus, was not so sure of ascertaining the length of the solar year (Ptol. *Almag.* III, 1, 1), and the astrologer Vettius Valens (IX, 11) *c.* AD 155 (*cf.* O. Neugebauer, *HTR* 1954, 65) still quoted several values exceeding 365 $\frac{1}{4}$ day (*cf.* O. Neugebauer, *Rivista degli studi orientali* 1949, 92).

Since Dunn considers a fraction of a day as a major error, his Athenian astrologoi were remarkable observers; for the sun at times reaches its greatest declinations at night.[27] See Goldstein and Bowen, "Introduction of Dated Observations ... in Greek Astronomy," *Archive for History of Exact Sciences* 43 (1991) 93 ff. The solar year is roughly 365 days 6 hours 9 minutes and 9.5 seconds with the lunar month on average 29 days 12 hours 24 minutes 3 seconds.[28]

Some ancient testimony about the solstice was presented in *ZPE* 128 (1999) 91–92, to which a few words may be added. Herodotos (2.109) wrote: πόλον μὲν γὰρ καὶ γνώμονα καὶ τὰ δυώδεκα μέρεα τῆς ἡμέρης παρὰ Βαβυλωνίων ἔμαθον οἱ Ἕλληνες. (E. Powell tr.: "Howbeit the Greeks learned the sundial and its pointer and the twelve parts of the day from the Babylonians"). Powell, "Greek Timekeeping," *CR* 54 (1940) 69–70, suggested the passage was an interpolation, but D. S. Robertson, "The Evidence for Greek Timekeeping," *Ibid.* 180–182, replied that the words ἡλίου τροπίου and polos were used by Aristophanes in the same sentence of a new fragment and provides excellent contemporary evidence for an Athenian sundial.[29] Gnomon was used by Hipparchos and the fifth-century astronomer Oinopides (cf. Von Fritz, *RE* s.v. [1937] 2259). In a passage often neglected, Theophrastos (*Sign.* I.4) tells us that Meton's teacher, Phaeinos, observed the solstices (τὰς τροπάς) from Mount Lykabettos, and two other astronomoi from other mountains. Stephanie West, "Cultural Interchange over a Water-clock," *CQ* 67 (1973) 61–64, refers to *P. Hibeh* 27, which shows

[27]For example, in 429, the solstice would take place after midnight from June 26 to 27; *BCH* 85 (1961) 31.

[28]Bonnie Blackburn and Leofranc Holford-Stevens, *The Oxford Companion to the Year* (1999).

[29]This fragment is not in Kock and Edmonds. Pollux (9.46) quotes a fragment from the *Gerytades* where a corrupt ἕκαστα is emended to εἶτα by Porson and translated by Edmonds (frg. 163), "This thing the pole? then bless my soul, tell me the time of day, please": πόλος τόδ' ἐστίν; εἶτα ποστὴν ἥλιος τέτραπται;

that one Greek about 300 B.C. referred to an instrument in Egypt for measuring time at night and defends the thesis of early Egyptian influence on the Greek calendar. As noted in *ZPE* 128 (1999) 92, the word στοιχεῖον is attested as being the shadow cast by the heliotropion and is used by Aristophanes; see LSJ, Montanari, and Aristophanes *Eccl.* 652; Euboulos, the comic poet, frg. 119.7; Lucian *Gallus* 9 and scholion. See also Photios s.v. στοιχεῖον: ἐκάλουν τὴν αὐτῶν σκιὰν ᾗ τὰς ὥρας ἐσκοποῦντο. The Souda (178 Δεκάπουν σκίαν) refers to the *Eccl.* passage and adds ('Αριστοφάνης) περὶ ἡλιοτροπίου φησί.

James Evans, *The History and Practice of Ancient Astronomy* (Oxford 1998) devotes Chapter 3 (pp. 129–161) to the subject of sundials. For a fourth-century stele identified as a heliotropion, see *SIG*³ 1264. H. L. Lorimer, *Homer and the Monuments* (1950) 80–84, comments on this heliotropion in connection with Homer's phrase ὅθι τροπαὶ ἠελίοιο in *Od.* 15.404. See the commentary of Hoekstra. In *IG* XI.2.287A line 117, the instrument has a fence around it.[30]

Meton.[31] Toomer, in *OCD*³ (1996) 381, writes:

> From the earliest times the Greeks, like many other peoples, named certain prominent stars and groups of stars. Homer speaks of the Pleiades, the Hyades, Orion, Boötes, the Bear ('also called the Wain', *Od.* 5. 273), and the 'Dog of Orion' (i.e. Sirius, *Il.* 22. 29). Hesiod mentions all of these, and uses their heliacal risings and settings to mark the seasons and times for agricultural operations (e.g. the rising of the Pleiades for harvesting, and their setting for ploughing, *Op.* 383–4). This traditional 'agricultural calendar' was elaborated and codified in the later 'astronomical calendars' of Meton, Euctemon and their successors.

[30]In Polybios 5.99.8, the word is used as a toponym, apparently a hill.

[31]For some bibliography on Meton, see *Thucydides' Pentekontaetia* (1995) 189–193.

O. Neugebauer, *A History of Ancient Mathematical Astronomy* 2 (1975) 622, wrote, "One realizes that Meton did not attempt to introduce a new lunar calendar but intended to establish a definite starting point in the solar year for the construction of parapegmata."

In the digest of solar calendars with zodiac months collected in the Parapegma, or calendar, of Geminos (attributed to Geminos, but much older), Meton is named but once (Manitius, p. 212), on the rising of Sirius (Day 25 of Cancer), and that two days different from the date given by Euktemon (Day 27), who, by contrast, is cited 47 times. Eudoxos, who is said to have lived in Egypt and wrote his *Octaëteris* there (Diogenes L. 8.86 ff., Plutarch *Mor.* 359C, Strabo 17.1.29), is cited 60 times. Different parapegmata had different epochs; that in Geminos was Cancer (Egyptian influence ?). The parapegma of Eudoxos commenced with the Lion, that of Aratos and Kallippos with Cancer or Capricorn, that of Poseidonios with Ram; see A. Bouché-Leclercq, *L'astrologie grecque* (1899) 129. The basic study of Parapegmata is that of A. Rehm, but see also the important work of R. Böker, *RE* Suppl. 9 (1962) s.v. Wetterzeichen, 1609–1691. For the dates of the fixed stars named in the parapegmata, see cols. 2427–2430 in Boll's article on Fixsterne in *RE* (1909). The identification of fixed stars and star groupings must have been prior to the parapegma tradition.[32]

[32]Just as there were different epochs for solar parapegmata, so scholars believe there were different epochs for the lunar calendars of Greek city-states. Rehm, followed by Samuel, *GRC* 114, concluded that Miletos before Alexander began the year with the autumnal equinox. Pédech, *RSA* 2 (1972) 6, thought the Boiotian year began at the full moon after the winter solstice; see Tuplin, *Klio* 69 (1987) 83. Meiggs/Lewis, *GHI* p. 171, have the winter solstice for Delos. That the parapegmata of Euktemon and others had any influence on the calendars of Greek states, including Athens, is completely fallacious.

As presented in *BCH* 85 (1961) 39, following Rehm, Euktemon codified a zodiacal calendar of 365 days, running from solstice to solstice. Meton/Euktemon were not the first astrologoi to know that the solar year approximated 365/6 days. Meton/Euktemon were not Einsteins creating a new theory about solar phenomena. Some modern authors in their desire to get Julian dates write as if Meton assembled Attic farmers in Skirophorion 432 and presented them with a hitherto unknown solar calendar, which they proceeded to inscribe on stone. To associate this codification with the creation of a new prytany calendar lacks all substance, which is true of Dunn's papers, who writes, "Meton's parapegma and his heliotrope made possible the new solar calendar adopted by the council in the late fifth century."[33]

* * * * * * * *

Below is a collection of some fifth-century texts which have been invoked for determining correspondences among prytany, festival, and Julian calendars. Dinsmoor produced an 'ideal' calendar of months which Meritt reproduced throughout his studies as recently as his *Athenian Financial Documents*, the conclusions in which are not as secure as he posited, although they have been followed in *HCT* IV.

1. The earliest inscription which provides even a remotely approximate calendar equation is *IG* I³ 364, which gives two payments for the expenses of the squadrons sent to Kerkyra in 433 B.C. The Panathenaia (Hekatombaion [I] 28) falls between the two payments, since two different treasurers of Athena made the payments. The first was on Prytany I.13 (τρε̃ς καὶ δέκα ἑμέραι ἐσ-

[33]*Ancient World* 29 (1998) 45. Dunn claims that Aristophanes in the *Nubes* was complaining about the introduction of a new *prytany* calendar! Aristophanes refers only to Selene, not to Helios. In *Aves* 992–1020, Aristophanes presented Meton, not as an astrologos, but as a γεωμέτρης.

ελελυθυίαι ἔσαν), the second on τε̑ι τελευταίαι ἑμέραι, which is Prytany I.37, if we apply the Aristotelian pattern to the fifth century (*Calendars of Athens* p. 96). Some historians date the battle in August: Meiggs and Lewis, no. 61; Lewis, *CAH* V^2 (1992) 375; Hornblower, *CT* I p. 67. Gomme (*HCT* 1.196–197) devotes a page to the chronology, concluding, "Sybota, therefore, is to be dated end of August or early September 433; almost all modern scholars agree on this."[34] The second squadron of twenty ships arrived late in the day after the battle, in which the first squadron did not engage, was over: Thucydides 1.50–51. According to the tables of Parker and Dubberstein, in Babylonian chronology the first new moons after the solstice in 433 were 6/28, 7/28, 8/26.[35] Hekatombaion (I) 28 must fall within the interval of 24 days between the two payments, and Prytany I.13 cannot be earlier than Hekatombaion 1 of the archonship of Apseudes (433/2). The payments need not represent the date of departure, but must be within a few days of it. Allowance of a week or more to make the voyage would seem to be reasonable.[36] If we assume, for example, that Hekatombaion 1 = 6/28 and test the theory that Prytany I.1

[34]Cf. G. Busolt, *Gr. Gesch.* III.2.783–786 ("in der ersten Hälfte des September"); Hammond, *JHS* 65 (1945) 31 (Late August or early September); *CAH* V. Chronological Table 1 (September); Meritt et al., *ATL* 3.319–320 (September).

[35]Cf. the tables (after Ginzel) for astronomical new moons in mean Greenwich time in E. J. Bickerman, *Chronology of the Ancient World*, 1st 1968 ed. p. 116. The second edition omits the tables.

[36]Cf. W. W. Tarn, "Fleet-Speeds: A Reply to Dr. Grundy," *CR* 23 (1909) 184–186. In J. S. Morrison and R. T. Williams, *Greek Oared Ships* (Cambridge 1968) 309, the only specific clue they give to the speed of the *trieres* is, "Xenophon says that the voyage under oar from Byzantium to Heracleia about 140 miles east along Black Sea coast of Asia Minor took 'a long day'." The speed of fleets is discussed by Lionel Casson, *Ships and Seamanship in the Ancient World* (Princeton 1971) 292–296.

fell on the summer solstice of June 27,[37] we have Prytany I.13 = July 9 and Prytany I.37 = August 2. The test fails, if we assume the battle occurred in late August. If we assume Hekatombaion 1 fell on 7/28, the theory of the summer solstice fails also. The chronology preceding the battle and Prytany I.1 might be restudied for an earlier date in August.

2. 432 B.C. Milesian Parapegma.

> [θ]ερινῆς τρο[π]ῆς [γε]-
> νομένης ἐπὶ ᾿Αψεύδους
> Σκιροφοριῶνος ΙΓ̅, ἥ-
> τις ἦν κατὰ τοὺς Αἰγυ-
> πτίους μία καὶ Κ̅
> [τ]οῦ Φαμενώθ, ἕως
> [τῆ]ς γενομένης ἐπὶ
> [Πολ]υκλείτου Σκι-
> [ροφορι]ῶνος ΙΔ̅, κα-
> [τὰ δὲ τοὺ]ς Αἰγυπτί-
> [ους τοῦ Παυ]νὶ τῆς ΙΑ̅,
> [κατὰ δὲ τὸ Μιλ]ήσιον

See also Diodoros 12.36; Ptolemy *Almagest* 1 p. 205 (ed. Heidberg). The epoch date for the astronomical cycle of the parapegma falls on Skirophorion (XII) 13 (= summer solstice) of this same year of Apseudes in No. 1. Dinsmoor, *Archons of Athens* 327, adopted the equation Skirophorion 13 = Prytany I.1 in 432 B.C. as the springboard for his reconstructed calendar,[38] criticizing Meritt for neglecting to do so; cf. *BCH* 81 (1957) 296. Toomer, whose articles in *OCD*^{3} (1996) include Meton, Euctemon, Callippus, Hipparchus, Astronomy, and Astronomical In-

[37]I follow Tafel V in F. K. Ginzel, *Handbuch* II p. 578. M. L. West, *Hesiod Works and Days* (1978) 381, following P. V. Neugebauer, says the summer solstice fell a day later (June 22/23) in antiquity than now.

[38]Dinsmoor reaffirmed his position in *The Athenian Archon List* 211.

struments, wrote in *Dictionary of Scientific Biography* IX (1974) 338:

> The equations with the Egyptian calendar plus the archon years enable us to determine these dates as 27 June 432 B.C. and 26 June 109 B.C., respectively. They are exactly seventeen nineteen-year periods apart; and one can deduce, first, that the second date is not an observed solstice but one computed from the first by means of a fixed calendrical scheme and, second, that the Skirophorion dates must have been taken not from the Athenian civil calendar but from an artificial astronomical calendar. The first is the starting date of Meton's first cycle. The second must be the date in the current Callippic cycle, for the interval corresponds to 323 years of 365 $\frac{1}{4}$ days. The transition from Metonic to Callippic reckoning may also explain the shift of one day (from 13 to 14 Skirophorion) of the date of the solstice after an integer number of nineteen-year cycles.

Apparently, the 432 dates in the parapegma must have survived a long tradition. One assumes that there was no error in transmission. Scholars concede that the Egyptian part of the equation was a calculated date. Leo Depuydt, *Ancient Society* 27 (1996) 28–45, says that it was calculated by counting two Kallippic cycles back from the solstice in -279. Earlier, A. C. Bowen and B. R. Goldstein in *A Scientific Humanist: Studies in Memory of Abraham Sachs* (1988) 39–80, suggested that Hipparchos counted backwards from the vernal equinox of 1 Phamenoth in -127. One difficulty is that Skirophorion 13 was ahead of a real lunar calendar. As discussed most recently in *Thucydides' Pentekontaetia* (1995) 187–194,[39] astronomers used the names of months in the Athe-

[39]Those who speak of Skirophorion 13 as an observed date by Meton can hardly mean that he observed the heavens through the night, and in the early morning (πρωίας, Ptolemy) found that it was then at its farthest point in a northerly direction. However the date was determined, it must derive ultimately from a series of observations. Meton's astronomical observations are

nian archon's calendar for months of the 'artificial' calendars they reconstructed. Such calendars used fixed intercalary months, which was not always true of the Athenian calendar; so the months in astronomers' calendars did not duplicate those in he festival calendar at Athens, which, moreover, was subject to considerable tampering. As reported in *ZPE* 128 (1999) 79, the preliminary announcement that an archon-table will be offered with a fixed 19-year cycle will receive considerable scrutiny with revived interest in the subject, but any application to the fifth century runs counter to all tables so far published.[40]

Concerning the Diodoros passage (12.36.3). O. Neugebauer (*HAMA* 2.617) writes, "When Diodorus says that the Metonic cycle was still in use in his time (i.e. around 50 B.C.), 'with the majority of Greeks' he obviously did not expect his readers to think of the civil calendars and refers explicitly to Meton's success in predicting the 'significant' days for the weather prediction from stars." Students of the Athenian festival calendar have erred badly and continue to err in assuming that the "cycles" and astronomical calendars in the literature, particularly Geminos, have any relation to civil calendars.

3. The Logistai Inscription: *IG* I[2] 324; *IG* I[3] 369 (an unsatisfactory text and commentary). Meritt offered texts in *Athenian Calendar* plate 1, and *Athenian Financial Documents* plate XII, as summarized in pages 52–53 of the former publication. This inscription spans the years 426/5 through 423/2, recording 17 payments with the interest in terms of the prytany calendar, as well as

---

the earliest that Ptolemy thought worth mentioning, and we are told that Meton erected an instrument for his observations. Goldstein and Bowen study instruments in *Archive for History of Exact Sciences* 43 (1991) 93–132. See also Toomer, *OCD*[3] s.v. For parapegmata, see E. C. Dorda, "Arato y el Parapegma," *Parola del Passato* 286 (1996) 5–18.

[40]In Chapter VI (52–73) of *The Choiseul Marble*, I offered a survey of theories about the Metonic cycle.

yearly totals and a grand total. The payments were loans made to the Athenian State from sacred treasuries. The record of the Other Gods gives what can be restored for two months in the festival calendar of 423/2.

Text. The text is not preserved within six letter-spaces of the right margin nor within four of the left. R. P. Austin, *Stoichedon Style* (1938) 60, commented at length on the puzzling irregularities in Meritt's reconstruction.[41] I catalogued these many anomalies in *Athenian Calendars on Stone* (= *CPCA* IV.4 [1963]) 310. In the first text Meritt approved with M. Lang in *Hesperia* 39 (1964) 146–167, their text has nineteen examples of the printer's inverted slug, used as a sigla for damaged letter-spaces in the restored parts. Their claim is that the mason inscribed a stone which had flaws, not only at the right edge but in restored spaces in the center of the stone, although there are none in the preserved part.[42] In my first inspection of the stone in Athens, I realized that some of Meritt's readings were questionable. Seeking independent appraisals, I published in *Ancient Athenian Calendars on Stone* 271–272 with photograph on pl. 20a, a written communi-

[41]Quoted in *Calendars of Athens* (1947) 100 n. 13 and *AJP* 85 (1964) 46–47.

[42]See *CSCA* 1 (1968) 215. The Lang-Meritt text in *Hesperia* 33 (1964) 146–167 assumes (1) "damage" at the lost right edge of the stone for eleven lines, thus eradicating sixteen letter-spaces and introducing into the Leyden system a new siglum, the inverted slug; (2) "damage" of one space within line 51; (3) the abacus operator made a mistake of 3610 dr. for 5610 on the stone (pp. 155–156); (4) for the sum 2610, the abacus operator computed interest only on the 610 dr. (pp. 154–156); (5) "three obol pebbles were put mistakenly" in the drachmai column (155–156); (6) the stonemason inscribed the wrong figure (p. 154). See *Hesperia* 34 (1965) 135–136. Moreover, I have questioned that the table on which they made their calculation was an abacus rather than a gaming table, particularly since Stroud found one on the akropolis rock at Agios Andreas; see "'Five Lines' and *IG* I^2 324," *CSCA* 1 (1968) 187–216. Later, a table was found at Corinth marked demosion, possibly used in some stoa.

cation from J. Caskey including a diagram and reports of inspections of the stone by Vanderpool, Mitsos, Mrs. Caskey, E. Harrison, H. Caroll, confirming my doubts about the readings for numerals in line 32.[43] S ee also *CSCA* 1 (1968) 204. In *Hesperia* 34 (1965) 237–238, M. Lang replied, "Of the six scholars cited by Pritchett ... only two are epigraphists and the drawing of the area in question ... was not made by either of these two." In an attempt to conclude the debate, I published in *Phoenix* 23 (1969) pls. 2 and 3, a series of twelve photographs, taken by Emil, a well-known professional photographer in Athens, from all angles. All of this bibliography and more was ignored by D. Lewis in *IG* I³ 369, even omitting any reference to photographs in his critical apparatus, who reads Meritt's text undotted. Meritt's sixth text, this with M. Lang, appears in *CQ* 62 (1968) 84–94.

[43]Whenever I have crossed swords with the Meritt/McGregor/Lewis team, *IG* I³ has always followed Meritt. In line 32, where we have at the edge of the stone the end of a series of numerals, there are traces of the upper parts of three uprights. I published not only the reports of the six scholars but a detailed sketch and a photograph of a latex squeeze (pl. 20A). Lewis mentions none of this, but adopts Meritt's text (undotted), commenting, "de lectione non dubitamus." Again, critical for the restoration of the name of the secretary and the Meritt-McGregor equation of Prytany I.1 = Hekatombaion I.1 for 422 B.C. in *IG* I² 220 (= I³ 285), I published, after much debate, in *REG* 86 (1973) 35–44 ("*IG* I², 220: Affidavits") both the affidavits of G. Daux, B. Helly, and T. Drew-Bear and four detailed photographs taken at all angles. Moreover, I reported that when McGregor was in the Epigraphical Museum, I invited him to show me and others present Meritt's letters on the stone. He could not, but claimed they were on a Princeton squeeze. To omit the names of such eminent epigraphists from the *IG* I³ commentary and to publish an erroneous text is in character with this corpus. Again *IG* I³ omits all references to photographs of the top of the lapis primus (I³ 259) and the so-called postscript, as in *GRBS* 7 (1966) pls. 1–9 facing p. 126 and in *CSCA* 5 (1972) pls. 1–4 facing p. 154. The photographs are far more valuable to the reader than a highly prejudiced commentary. The practice in *IG* I³ is counter to that in all other corpora in the history of Greek epigraphy.

Uncontested evidence about the prytany calendar in *IG* I^2 324 proved: (1) the four years contained 1464 days, or 366 × 4; (2) the six payments of the first year indicate that that year was of 366 days. These payments are so distributed that one may postulate that the first six prytanies were of 37 days and the last four of 36 days; (3) the first three prytanies of the second year contained 37 days each; (4) the last prytany of the fourth year was of 36 days. These were also Meritt's determinations, and were so adopted in *Calendars of Athens* 95–96.

The restored portion has given rise to controversy. The interest per diem was 1/4 obol, the smallest monetary unit used, on 1250 drachmai, or 1 drachma on 5 talents. Where the principal is preserved, one may compute the interest, but we have a number of cases where only the interest is preserved in whole or in part, which has given rise to different computations, one which denied four years of 366 days and assumed errors and irregularities by the abacus operator and stonemason, examined by me, "Gaming Tables and I.G., I^2, 324," *Hesperia* 34 (1965) 131–147. I know of no further research which denies the 366-day prytany calendar, although other problems are hotly contested.

Meritt and McGregor, *Phoenix* 21 (1967) 88, presented the following table for the dates of Prytany I.1 in the period 426/5 – 422/1:

	FESTIVAL YEAR			CONCILIAR YEAR			
	Attic inter-calation	Date of Hekatom-baion 1	Number of days	Date of Prytany I, 1			Number of days
426/5	(1)	July 11	384	Skir. 23/24	=	July 4	366
425/4	0	July 29	355	Skir. 5/6	=	July 4	366
424/3	0	July 19	355	Skir. 16/17	=	July 5	366
423/2	0	July 9	354	Skir. 27/28	=	July 6	366
422/1	0	June 28	354	Hek. 10	=	July 7	366[44]

[44]Cf. Meritt's table in *AFD* 176.

Here Prytany I.1 is seven or eight days removed from the solstice on June 27/28 (Ginzel).

My computations with accompanying tables in *Calendars of Athens* 103; *Ancient Athenian Calendars on Stone* 290–312; and *Choiseul Marble* 100–101, are based on the assumption that the record was not what W. S. Ferguson referred to as a "virtual falsification of the record," i.e. a theory that payments were made by officials not in office.[45] Thus, I believe that in 426/5, the payment made on Prytany II.4 must be after Hekatombaion 28 of 426, that on Prytany I.12 of 423/2 must be after Hekatombaion 28, etc. These and other dates disproved the Meritt-McGregor theory of four ordinary years in succession.

An additional comment is in order about the names of the two months in the inscription, both restorations, particularly since Dover in *HCT* IV 268 makes one a fixed date in his calendar.

In lines 58 and 79, we have two equations in which only five letters of seventy-five are preserved in line 58 and four letters of the day of the month in line 79. Meritt reads [hεκατομβαιô]νος and [Σκιροφοριôνος]. The letters in line 79 continue to be misread by Meritt/Lang/Lewis, as demonstrated by the photograph of a latex squeeze in *AJP* 85 (1964) 40–50. In *The Choiseul Marble* (1970) 100–101, I attempted to show that the restoration of Metageitnion in line 58 with the same number of letter-spaces would require at least one intercalary year in Meritt's period of four ordinary ones.[46]

Neither the right nor the left margin is preserved. The critical matter is the number of letter-spaces to be allowed for each line. Meritt throughout, followed by Lewis in *IG* I³, claims that lines 1–

[45]See "The Hellenotamiai and Athenian Finance," *Hesperia* 26 (1977) 295–306, particularly p. 304.

[46]In line 79, Ἑκατομβαιôνος (without the aspirate as in some cases) may be restored; see *Choiseul Marble* 101 no. 6 and *CSCA* 1 (1968) 213. Dover, *HCT* 4.268, follows Meritt.

74 were "stoichedon 75", and that below line 74 there were never more than 75 letter-spaces (Meritt).[47] R. P. Austin, *Stoichedon Style* (1938) 59–60, not noted in Lewis' bibliography, devoted a page to what he called Meritt's "puzzling irregularities" in the number of letters per line. In the upper part, Meritt prints one to six vacats at the end of lines and one to six in the restored parts. Below line 74, we cannot even attempt restorations for lines 80–94 except at 85/86 where Meritt and Lewis read 'Αφ[ροδίτε|ς] as above in 69/70 they read [Κυνοσάργε|ι] and in 8/9 [ᵛτ|όκος]. It is pretense to claim that we can be sure that there were 14 letter-spaces in lines 58 and 79 for the names of the months and what these names were. The use of the spiritus asper varies. Whereas the month in line 79 was in Prytany X, the prytany and its day in line 58, which preserves only 5 letters, are entirely restoration. The degree of restoration is illustrated by the facsimile drawing in Plate 12 of Meritt's *Athenian Financial Documents*.

Focusing on Aristotle's statement (*Meteor.* 1.343b) that a comet was seen at Athens in Gamelion (VII) 427/6 about the time of the winter solstice,[48] Lotze (*Philologus* 111 [1967] 39–40) offers a sequence of months for the year different from that of Meritt. A more important astronomical feature to consider in the period of our inscription is recorded in the Venetus scholia on Aristophanes *Clouds* 584 that an eclipse of the moon took place in Boedromion (III) of 425/4: ἔκλειψις ἐγένετο Σελήνης τῷ προτέρῳ ἔτει ἐπὶ Στρατοκλέους Βοηδρομιῶνι. The date was October 9: Boll, *RE* s.v. Finsternisse (1909) 2355. If the calendar was regular, this full moon in the third month (30 + 29 + 15 = 74) would yield a calendar with Hekatombaion (I) 1 falling on July 28 (9 + 30 [Sept.] + 31 [Aug.] + 4 = 74). This accords well with

[47] Lewis even reports lines 74–110 as stoichedon 74.

[48] According to Gundel, *RE* s.v. Kometen (1921) 1183, this comet was seen in January in the northern sky.

the table of new moons for Babylon in R. A. Parker and W. Dubberstein, *Babylonian Chronology* 31, which shows a new moon on 7/29. This date was the second new moon after the solstice. Such a date nullifies any theory of a Metonic cycle.

4. Thucydides 4.118.12, 4.119.1, and V.19.1.

> 424/3 Elaphebolion (IX) 14 = Gerastios 12 (Sparta)
> 422/1 Elaphebolion (IX) 25 = Artemision 27 (Sparta)

In 4.118.12, Thucydides says the truce began in Athens on Elaphebolion (IX) 14 and in 4.119.1 that it was ratified at Sparta on Gerastios 12. Scholars generally assume that the two days correspond, implying that the date was fixed on for this commencement, but A. E. Samuel, *Greek and Roman Chronology* 93, writes, "The implication of Thucydides is that the Spartans swore some time after the beginning of the truce in Athens, suggesting that Gerastios 12 was later than Elaphebolion 14." However, 4.119.1 is part of the text of the truce, enclosed in quotation marks by some editors, which was ratified in Athens by Spartan ambassadors. Many editors follow Kirchhoff in deleting καὶ ὤμοσαν.[49] An interesting feature is that this truce was consummated during a day now normally assigned to the City Dionysia. The fact that the Spartan months Artemision and Gerastios are both associated with Elaphebolion proves that the sequence of ordinary and intercalary years was not identical in the two calendars. Moreover, in 423 the day in the Spartan month was two days earlier than that in the Athenian month, but in 421 two days later; so the νουμηνία's were not the same. On each of the two occasions either the Athenian or the Spartan calendar, or indeed both, were out of step with the moon. At Athens, Elaphebolion was a not uncommon month for tampering with the calendar. We know next to nothing about the Spartan calendar. Errors in numerals in Thucydides, as well as in

[49]Also Hornblower *CT* II p. 371.

Greek texts in general, were studied in *Thucydides Pentekontaetia* 24–61, but none have been proposed for our equations.[50]

Not all editors are agreed with the above. A. W. Spratt, in his commentary, wrote, "Most editors agree in regarding the treaty as drafted at Sparta and sent to Athens for approval: from this point of view sections 1–10 will all be a Spartan draft. This view Poppo regards as confirmed by the words of 9 εἰ δέ τι ... διδάσκετε." Haacke, Dobree, and Keil offer different divisions of the terms. C. E. Graves writes, "It appears to have been drawn up and agreed to at Sparta, and then sent to Athens for ratification." I agree about sections 1–10. Arnold suggests that in the case of the truce the Gerastios date is that which the Spartan government swore at Sparta before ratification at Athens and in the case of the Elaphebolion date for the peace in 5.19.1 that of the Athenians at Athens before it was ratified at Sparta.

5. *IG* I³ 78. Unknown Year. Decree concerning the Eleusinian First-fruits.

An amendment by Lampon (lines 47 ff.) of a decree passed before the ninth prytany (60) prescribes that the archon of the next year shall intercalate a second Hekatombaion,[51] which requires an interval of a lengthy period. One might think that the year of the decree was ordinary to avoid two intercalary years in a row, but the concern of the ekklesia was to extend the interval of days before the next Eleusinian festival and the nature of the current year would have been immaterial. The following year may have been projected as ordinary. The assumption by calendarists that the year of the decree was ordinary is not secure.

[50]I should have mentioned that in 7.16.2 the correct numeral was read by only one manuscript (H) and Valla.

[51]For the fifth century, the only attested case of an intercalary month is this Hekatombaion. The choice of months was at the caprice of the demos. Cf. Lotze, *Philologus* 111 (1967) 41. See Table 3 in Chapter 1.

Dittenberger in *SIG*^{3} 83 and Hiller in *IG* I^2 76 dated the decree in "c.a. 423/2". M. N. Tod, *GHI*^{2} no. 74, with additional bibliography on p. 264, wrote, "Unfortunately, there is nothing in this measure to determine its date, and the suggestions of modern scholars range from 445 to 417 or even 416 B.C.", favoring with misgiving a date in the spring of 418. Meiggs and Lewis, *GHI* no. 73, devoted more than a page to options for the date, suggesting in the heading "? c. 422 B.C.", and writing, "Dinsmoor's date can certainly be made to fit, but his key argument that Kekropis held the eighth prytany both in this decree and in 416–5, relies on restorations of No. 77, ll. 47, 49, which are by no means certain. We should not be in the least surprised if new discoveries assigned the decree to, say, 435." Dinsmoor, Meritt, Meritt and McGregor favored a date in 416/5,[52] and have been followed by B. Smarczyk (*SEG* 42.17). O. Hansen, *Eirene* 27 (1990) 47–48, proposes 421/0, attributing the delay in the calendar to the purpose of repairing buildings damaged in war. Lewis, who usually follows Meritt slavishly, in *IG* I^3 78 (1981), gives the date as "c.a. 422 ?". He notes that Mattingly favors 425/4. C. W. Fornara, *Archaic Times to the End of the Peloponnesian War* (1977) no. 140, gives an English translation. I lean to the conclusion of Meiggs, *The Athenian Empire* (1972) 304, "One of the last years of the Archidamian war provides a rather more convincing setting."

6. Aristotle *Ath. Pol.* 32.1 (412 B.C. Loeb tr.):

> XXXII. This then was the constitution drawn up by the Hundred elected by the Five Thousand. These proposals were carried by the multitude, being put to the vote by Aristomachus, and the Council in Callias's year was dissolved on the 14th day of the month of Thargelion before it had completed its term of office; while the Four Hundred came into office on Thargelion the 21st <sic, 22>; and the Council elected by lot was due to enter office on

[52]See *Proc. Amer. Phil. Soc.* 115 (1971) 109.

> Scirophorion the 14th. In this way therefore the oligarchy was set up, in the archonship of Callias.

This equation Skirophorion (XII) 14 of the archontic year 412 = Prytany I.1 was rejected by Dinsmoor, *Archons of Athens* 328–330 with the claim that Aristotle miscalculated or that the equation should apply to 413 B.C. In his *Athenian Archon List* 211, he reaffirmed his position that it was an error. At least, he raised the question as to how Aristotle would know, what seems irrelevant, the date when a boule, which never came into existence or ratified any document, would have come into office. Meritt devoted the first two pages of an article on the hollow month in *Mnem.* 30 (1977) 217–218 to the argument that Aristotle computed the Skirophorion 14 date, and hence the months were determined in advance by rule; the Athenians had a schematic festival calendar. Meritt does not attempt to enlighten us as to the source of the other dates. Since this section does not agree with Thucydides, the source of the entire passage has been much discussed.[53] Andrewes, *HCT* V (1981) 250, says "30–32.1 offer few land-holds." Rather than taking the source as Androtion, apparently Jacoby, he writes, "I would take 30–32.1 to be based on a distinct document, preserved in private hands rather than published by Androtion." Rhodes in his commentary takes 32.1 as pointing "to an *Atthis* rather than a work of polemic" (p. 367) and suggests that Aristotle "used a work not known to us" (p. 368). After debating the archival theory in *PCPS* 202 (1976) 23, Andrewes wrote in *HCT* V (1981) 235–236:

> The proclamation would give only one date, 21/22 Thargelion, and the source of the other two must be sought elsewhere. They may come from the same source, between them defining the period for which the term of the democratic Council had been cut short,

[53]Cf. Mathieu on p. viii of the Budé edition on the matter of authenticity.

> thus accounting for the irrelevant 14 Skirophorion. The latter, as argued above, should come from a contemporary source, and a speech from one of the many trials of the period, perhaps Antiphon's famous speech, is more likely than a document of the time ferreted out by later research. An orator will sometimes give dates by the lunar calendar, as D. xix. 57–60, or in a non-political case Antiphon vi.42–4, and we have to imagine this orator making a point which depended on the length of time by which the deposed Council had been cut short: that might e.g. be a reference to the length of time for which they had been paid (Th. 69.4). We should then have to suppose that Aristotle, or perhaps his source, combined this datum with what he extracted from the proclamation of the Four Hundred, thereby producing the anomaly of a week in which Athens had no Council. Certainty is naturally impossible, but this relatively economical hypothesis would account for the otherwise puzzling fact that Aristotle bothered to mention the date at which a new democratic Council would have entered office but for the intervention of the oligarchs.

Andrewes continues by rejecting the idea that the equation was deduced from any documents in the archives.[54] J. J. Keaney, *The Composition of Aristotle's Athenaion Politeia* (Oxford 1992) discusses the sources for many passages, but nothing specific on 32.1. On p. 97, he comments on the chiastic structure in the calendar references. Elsewhere, he has reminded us of the use of alphabetical numerals, and one might wonder whether the second "14" is correct. In *CAH* V² (1992) 475, Andrewes wrote, "The mainly narrative sequence, 29 and 32.2–3, not too distant from Thucydides' version but tending to exculpate the reformers, may well be based on the *Atthis* of Androtion," which he says was the

[54]In the record in Thucydides, the historian records the seizure of the bouleuterion in 8.69.4–90.1. At the time, Agis was at Dekeleia and came down to the very walls of Athens (8.71.1). The Athenian fleet was at Samos (ἔχοντες τὰς ναῦς, 8.76.4) with 82 ships (8.79.2), with reinforcements which brought the total to 108 (8.79.6).

obvious up-to-date source when *Ath. Pol.* was written. As suggested in *ZPE* 49 (1982) 257–258, the boule of 400, which came into power in Thargelion, must have suspended the 366-day prytany calendar. It no longer met by prytanies. One document we have from this boule, *IG* I^3 373, gives only a festival, or lunar, date in Hekatombaion, whereas all the other records of the treasurers of Athena date by the prytany calendar.

This was Meritt's prime example for a schematic festival calendar, which he submitted without any reference to bibliography on the problem. It seems to me quite probable that in the complicated machinery of government some non-legislative officials were allowed to fill out their terms.

7. Antiphon VI. *Choreutes* 44–45 (Loeb tr.).

> Moreover, after the present Basileus had come into office, there were thirty clear days from the first of Hecatombaeon onwards, on any of which they could have registered their charge, had they wanted to; yet they did not do so. Similarly, they could have registered it any day they liked from the first of Metageitnion onwards. But even then they did not do so: they let twenty days of this second month go by as well. Thus the total number of days in the present archonship on which they could have registered their charge, but failed to do so, was over fifty. ... I acted as Prytanis for the whole of the first Prytany save two days.

In his Introduction to the *Choreutes* speech in the Loeb, Maidment (pp. 236–237) deduced that the political situation favored a date in 412,[55] but that Meritt's treatment of the calendar (*The*

[55]R. C. Jebb, *The Attic Orators* 1 (1876) 63, citing Sauppe, wrote, "Its date may probably be placed soon after the Sicilian disaster." L. Gernet, in the Budé edition p. 141, writes, "La date plus précise de 412 est même probable (cf. Wilamowitz-Möllendorff, *Arist. u. Athen*, II, p. 347)." Dobson, in *OCD* 1st and 2nd editions, would also date the speech to c. 412. Kirchner, *PA* 1304, cites Wilamowitz as dating the speech "sub 421".

Athenian Calendar [1928] 121–122) proved that it was delivered in 419. He summarized Meritt's position as follows:

> §§ 44–45 of the *Choreutes* contain two indications of time: (*a*) the speaker was formally accused of φόνος before the Basileus on the fifty-first day after the latter took up office, *i.e.* on the 21st of Metageitnion, the second month of the Attic year: (*b*) he was Prytanis for the whole of the first Prytany of the year save two days (the implication being that he was forced to resign before completing his term of office, because of the charge of φόνος lodged against him with the Basileus). Now as in the latter part of the fifth century the opening Prytanies of a given year are known to have consisted of 37 days, the speaker must have held office for 35; and this gives the equation: Prytany I.35 = Metageitnion 21: or, in other words, Prytany I.1 = Hecatombaeon 16. The Council therefore assumed office in this particular year sixteen days after the commencement of the civil year ... the epigraphical evidence makes it clear that only 419 B.C. will meet the case. This is admittedly unsatisfactory in view of the marked difference in style between the *Choreutes* and *Herodes*; but the systematic variations in the Attic calendar have been convincingly demonstrated from the plentiful data available, and unless we are inclined to resort to arbitrary alteration of the figures contained in the MSS. of the *Choreutes*, we must rest content with the earlier date.

Since we use Julian dates to correlate the two calendars, our calendric computations depend on the sequence of ordinary and intercalary years as well as the epoch date for any series, which means the relation of Hekatombaion (I) 1 to the solstice. When we turn to Meritt's calendar, reconstructed largely from the Logistai inscription, we find the following sequence of years 425/4 – 419/8: O O O I O O O. The year 419/8 began before the solstice. Since I do not believe that payments were made by officials not in office in the Logistai inscription, the sequence is incorrect, as noted above.

K. J. Dover, "The Chronology of Antiphon's Speeches," *CQ* 44 (1950) 44–60, would also date the *Choreutes* in 419, but according to a different chronology. His conclusion is in part (p. 60): "The first prytany began on Hekatombaion 15, 16, or 17. We know that the first prytany of 422/1 began on Hekatombaion 9 or 10, since Skirophorion 23 was the tenth prytany of 423/2" etc. he accepts Meritt's schematic calendar containing the equation Prytany I.1 = Hekatombaion 10 (422/1) and Meritt's arrangement of intercalary and ordinary years to get Meritt's date. In the record of the Other Gods in the Logistai inscription, where neither right nor left margins are preserved, and there are inexplicable variations in the script, two payments are made in terms of the festival calendar (lines 58 and 79).[56] In both cases, the names of the months are restorations. That of Skirophorion in line 79 depends on the fifth-century calendar we reconstruct for coordination with the prytany date X.20. See *CSCA* 1 (1968) 213, and *The Choiseul Marble* 100–101. Meritt's reconstruction (*Calendars* p. 118) of the years 423 – 419/8 is O I O O O.

M. Gagarin, in his 1997 edition of the speeches of Antiphon pp. 244–245, follows Meritt's calendar. The interesting feature is Meritt's deduction without documentation that the first prytany was "always thirty-seven days long," thereby implying a predetermined fixed calendar, a thesis which he subsequently denied in many publications after Neugebauer and I proposed it.[57] Since we do not know the sequence of festival years and the dates for

[56]In lines 58 and the preceding line, there are but five letter-spaces preserved.

[57]For Meritt, the length of the first prytany was here predetermined, but neither he nor such followers as Woodhead, Lewis, or Rhodes explain why the other prytanies were not. Again, they postulate a rigid festival calendar of alternating full and hollow months in which Skirophorion ultimoi are determined mechanically, but invoke a lunar date for Hekatombaion 1's whenever they want a Julian date.

Hekatombaion 1's, the date of this speech has not been solved by the calendar; cf. *BCH* 81 (1957) 298. In short, I reject Meritt's date based on checkpoints which give him (*PAPS* 115 [1971] 114) four intercalary years within five and eight ordinary years within nine.

8. 407/6. Choiseul Marble.

Fundamental to any study of the Athenian prytany calendar of the fifth century is the disputed text on the reverse side of the Choiseul Marble (*IG* I^2 304B; I^3 377). That on the obverse side is dated by the archon to 410/09 B.C. Both sides record payments of money from the treasury of Athena. On the obverse side, payments are recorded by the prytany calendar only, but on the reverse side they were inscribed in terms of both the festival and prytany calendars. There are 27 equations, spanning five prytanies and six months. In his text published in *Athenian Financial Documents* (Ann Arbor 1932) with photographs in plates 7–10 and a facsimile drawing in pl. 11, Meritt, after his study of the stone in the Louvre, misread (p. 122) the last line of the upper part (27 = his 92) as πὶ τῆς Ἐρεχθηίδος πρότης πρ[υ]τα[νε]υόσ[ης]. The first line of the lower part is quite clear as ἐπὶ τε̃ς Ἐρεχθεΐδος δευτέρας—. Accordingly, he endorsed the view that the mason began his text in the middle of the stone, continued inscribing until he reached the bottom of the stone (now lost), and then completed his text in the upper, then empty, portion of the same stele. I corrected Meritt's phrase with πρότης to πρότηι τῆς πρυτ[ανείας]. By measuring 6.5 cm. from the left edge of pl. 21 in *BCH* 88 (1964) and 2 cm. from the bottom, one sees the iota clearly. This correction of one letter allows the calendar dates to run in consecutive order from the top of the stone. The last equation on the upper part gives: Metageitnion 8 = Prytany II.1; the first on the lower part: Metageitnion 20 = Prytany II.13.

The fact that the two calendars with units of different lengths progress pari pasu from the upper part to the lower is to my mind compelling evidence that avoids the basic absurdity that a mason began a long text midway on the stone with so much empty space above. In the controversy over the text never has any parallel been cited in the history of epigraphy.

A second physical fact is that the Meritt/Lewis interpretation has the last line of the text as Line 27 with a record extending into the second month and the second prytany. Surely the mason must have been given a record to inscribe to the end of the year and would not have planned his text to make the reader take a leap into time and space.

A third physical factor entails an explanation of the space between lines 27 and 28. Lines 1–11 can be restored as stoichedon 73–74; inscribed in Attic script (τε̂ς, τρίτει, ἑλλενοταμίαις, etc., with occasional Ionicisms of gamma and eta). But lines 12–27 were inscribed entirely free-hand by a different mason using predominantly Ionic letter-forms with τῆς, τρίτηι, ἑλληνο-, etc., although this has never been acknowledged by Meritt and Lewis. He abandoned stoichedon order with lines of ninety-odd letter-spaces,[58] crowded some letters, spaced others widely, engraved adjacent letters of different height, and inscribed lines which are not parallel. I deduce that lines 12–27 were by an apprentice, the worst workmanship I have seen in comparable documents, and when the original mason took over to inscribe the lower part, he returned to a stoichedon order (of 74; the stele tapers) and spaced his text a short distance below that of his Ionic apprentice.[59]

[58]Lewis in *IG* I^3 377, for example, would restore line 15 with 99 letterspaces, 22 with 88, 23 with 101, and 24 with 96; all a far cry from stoichedon 73 or 74.

[59]S. V. Tracy, *Attic Letter-cutters of 229 to 86* (Berkeley 1990) 212–215, 126, 145, 148, reports inscriptions with different cutters, but apparently at different times.

The second sentence of the commentary to *IG* I^3 377 reads, "Nos lapide ipso non examinato omnia e photographiis praecellentibus, sicut Meritt in textibus recentioribus." The statement is unique in this volume. Meritt in *Athenian Financial Documents* (1932) 114–115 wrote:

> "The determination of the text of this inscription is extremely difficult because of the badly weathered condition of the surface of the marble, but I have given in the facsimile copy (Plate XI) those letters or traces of letters which I have been able to read from the stone. Through the courtesy of Les Archives Photographiques, I am also able to reproduce in Plates VII–X photographs of the reverse face of the stele, which may be used to some extent as a control in studying the facsimile plate. Many of the readings, however, were obtained after patient study of the stone itself under different conditions of light and shadow, and with the help of squeezes, charcoal wash, and plastic impressions. A definitive control in doubtful cases can be given only by careful inspection of the stone as now preserved in Paris."

The theme of Meritt's *Epigraphica Attica* (1940) was architectural epigraphy, the need to work from the stone. Our stone is like a text once described by R. S. Stroud (*Stone* 285 n. 35): "*IG*, II^2, 363, is like a sleight of hand artist—one minute you see a letter then you don't—then there is another stroke which looks as if it might belong to something else. The traces can be interpreted many different ways."

As to photography, G. Daux (*BCH* 93 [1969] 946) under the title of 'Trompeuses photographies' once wrote: "L'épigraphiste ne peut, ne doit jamais se fier entièrement à une photographie, et la contrôle par lecture directe, par l'estampage aussi permet seul une certitude." How do Meritt and Lewis know that my photographs are "praecellentes" when they have not compared them with the text on the stone?[60] Twice I urged over a period of more than a decade that Lewis and Meritt in their trips to Athens stop off in Paris and examine the stone in a public gallery in the Louvre before presenting a new version. One would think that the stone was a cobra. It seems truly remarkable that what G. Huxley referred to as the Princeton school of epigraphy and its dependencies could offer a dozen or so treatments of a text over a period of about fifteen years without once looking at a stone on public display. Meritt along the way has cited McGregor, Lang, Woodhead, and Lewis as confirming his readings, some of which he subsequently changed. Lewis devotes a full folio page (358) of *IG* I^3 377 to a critical apparatus in which he poses as a referee in his general preference for Meritt's readings, although he admits that he has not examined the stone. In *Stone* 288, I quoted "affidavits" from Pouilloux and Roesch about debated readings, but one will look in vain for the names of these eminent epigraphists in *IG* I^3 377's critical apparatus.[61]

There were originally incised about 2200 letters above the blank space and 1775 below (and above the break). I doubt that any single scholar could faithfully decipher all letters. In *BCH* 101 (1977) 7–42,[62] I reiterated that the center and rightmost part of the surface, particularly that incised by the Ionic mason, is severely worn and with so many scratches that I inferred under magnification, perhaps wrongly, that the stone was a palimpsest. I concluded that the patina from the incised letters had brought up

[60]The same question holds for Woodhead in *SEG* 21 (1965) no. 79 and 23 (1968) no. 36. Cf. *The Choiseul Marble* 22 n.3.

[61]Elsewhere, as reported above n. 43, *IG* I^3 preferred readings from a Princeton squeeze (without photograph) as against the testimony of Daux, Helly, Drew-Bear, and myself working from the stone. For I^3 377, Lewis inconsistently does not give the reader the benefit of what he might see on the Princeton squeezes and its professional photographs.

[62]This article was not included in the bibliography in *IG* I^3 377 (1981).

traces which could be seen when water was applied. The text of line 23 where I thought I brought up traces of ἕ̣ν̣[ηι] κ̣α̣[ι νέαι] is the one most in dispute.[63] If one measures 3.5 cm. from the bottom of pl. 21 and 2.3 cm. from the bottom of pl. 22 in *BCH* 88 (1964), one may place a ruler marking this line. By measuring 3.5 cm. from the bottom of pls. 7 and 8 in Meritt's *AFD*, one may bring up this same line. The coloration of the patina in part of this line is displayed on pl. 2 of *BCH* 101 (1977).

The prytany calendar of *IG* I^2 304A (= I^3 375) and the upper part of 304B agrees with the Aristotelian pattern given above p. 145. My equation Prytany X 33 = Skirophorion ultimo suggested that the prytany calendar of 366 days, beginning and ending independently of the festival calendar, was at this point in time changed to one which followed Aristotle's rule in the *Ath. Pol.*, but four dotted letters (ἕ̣ν̣[ηι] κ̣α̣[ι νέαι]) on this illegible surface offers too weak a case to build a theory on and I suspend judgment on the reading. Adjustment would have been made in the final prytany. It would of course have been a coincidence if the prytany calendar of 366 days had come to an end precisely on the day Skirophorion ultimo. The following year would start with Prytany I 1 = Hekatombaion (I) 1, with the first prytany of 36 days, as for an ordinary year under the Aristotelian pattern. Line 23 is followed by the equations Prytany I 20 = Hekatombaion (I) 20, Prytany II.1 = Metageitnion (II) 8, and Prytany II.13 = Metageitnion 20, which assures us that Prytany I of this year began on Hekatombaion 1. Moreover, enough of the text is preserved on the stone to require the assumption, regardless of whose

[63]The editors of *SEG* 30.22 (1980, published 1983) commented, "So many new readings and restorations of these texts have been published since *IG*, I^3 that we cannot publish a full apparatus criticus here." Meritt in *AE* 1978 95–108, presents a new text of *IG* I^3 377 which differs sharply from those in *IG* I^3 377 and those previously published by himself, all from my photographs, without consulting the stone.

text is followed, that there were embolimos/exairesimos days in the festival calendar. The Meritt/Lewis position that payments were made to officials not in office, what Meritt calls "clumsy book-keeping", will be treated anew by Loren J. Samons.[64] The crowning argument of Lewis that, because we have six phylai in office in 408/7 in reverse of the official order, the remainder must be, signifies nothing. Ferguson's position was, not that the complete order was allotted at the beginning of the year, but that each phyle was allotted just before the beginning of its term. Even if we knew the position of eight phylai, we could not deduce the position of the ninth. Lewis (*CR* 86 [22] 1972. 250) argues as if the authorities decided to abandon allotment at the beginning of this particular year and assign phylai in reverse of official order, a sequence for which he offers no parallel in Athenian history, and then the next year return to their regular practice. See above p. 133.

In summary, it should be recognized that in all of Athenian fifth-century inscriptions, there is not a single unrestored calendar equation, giving the day in terms of both the festival and prytany calendars, until we reach line 3 of *IG* I² 304B = *IG* I³ 377. On the literary side, the most secure date is the hypothetical one given in Aristotle *Ath. Pol.* 32.1 for Skirophorion (XII) 14 = Prytany I.1 of 412/1. There are approximations such as deduced from *IG* I² 295 = I³ 364; see *Calendars of Athens* 107–108. Dates in the modern literature have been obtained by hypothesizing a prytany calendar of 366 days, or whatever number one chooses, on one side of the equation and a festival calendar of some sequence of ordinary and intercalary years on the other side, always allowing for the possibility of embolimoi/exairesimoi days, which we know are found in the fifth century. Finally, one tries to get a Julian date for at least one Hekatombaion (I) 1 in relation to the summer

[64] See "Loans of Athena in 407 B.C.," *Anc. Soc.* 8 (1977) 33–47; "The Hellenotamiai and Athenian Finance," *Historia* 26 (1977) 295–306.

solstice. The several imponderables, including the term of office of officials and whether there was falsification of the financial records, would require a separate study.[65]

[65]See, for example, pp. 98 ff. of *The Choiseul Marble*; and n. 63.

VI

THE ATHENIAN PRYTANY CALENDAR AND EKKLESIAS

Studies concerning the Athenian Prytany Calendar, including its relation to the Athenian Festival Calendar, attracted the attention of many scholars throughout the nineteenth century, and with the discovery of many inscriptions in the excavations of the Athenian Agora, the pace has been intensified in this century. For example, two scholars alone have devoted at least eleven separate treatments to the problem of how often the ekklesia met.[1] In the penultimate one known to me, E. M. Harris,

[1]Bibliography of the two scholars on the subject includes the following: Mogens Herman Hansen, *Eisangelia* (Odense 1975) 51–57; "How Often Did the Ecclesia Meet?", *GRBS* 18 (1977) 43–70; "Εκκλεσία Σύγκλητος in Hellenistic Athens," *GRBS* 20 (1979) 149–156; with F. W. Mitchel, "The Number of Ecclesias in the Fourth Century B.C.," *SO* 59 (1984) 13–19; "How Often Did the Athenian Ecclesia Meet? A Reply," *GRBS* 28 (1987) 38–50; *The Athenian Assembly* (Oxford 1987) 20–32; *The Athenian Democracy* (Oxford 1991) 132–136; "Was the Athenian Ekklesia Convened According to the Festival Calendar or the Bouleutic Calendar?", *AJP* 114 (1993) 99–113; "160 Theses about Athenian Democracy," *Cl. et Med.* 48 (1997) 219–220. Edward M. Harris, "How Often Did the Athenian Assembly Meet?", *CQ* 36 (1986) 363–377; "When Did the Athenian Assembly Meet? Some New Evidence," *AJP* 112 (1991) 325–342. The subject is also treated by R. M. Errington, "Ἐκκλησία κυρία in Athens," *Chiron* 24 (1994) 135–160, and by P. J. Rhodes, "Ekklesia Kyria and the Schedule of Assemblies in Athens," *Chiron* 25 (1995) 187–198. Errington argues that the schedule of prytanies in *Ath. Pol.* 43.4–6 was not put into effect until 336/5. Rhodes argues successfully, I think, that the schedule was not restrictive. Items specified for one ekklesia might be discussed at other ekklesias. "Matters not among those specified for an ekklesia kyria could be discussed at an ekklesia kyria." One might also look at the different list in Pollux 8.95. In *GRBS* 23 (1982) 331–350, Hansen gives a useful catalogue of

"When Did the Athenian Assembly Meet? Some New Evidence," *AJP* 112 (1991) 325–341, writes, "The purpose of this article is to rescue from the rubbish heap a nugget of information about the Athenian Assembly found in a scholium to Dem. 24.20." The author concludes that the Athenians convened three ekklesias per festival month rather than four per prytany, as stated by Aristotle *Ath. Pol.* 43.3. In his reply, "Was the Athenian Ekklesia Convened According to the Festival Calendar or the Bouleutic Calendar?", Mogens Herman Hansen, *AJP* 114 (1993) 112, while defending Aristotle for part of the fourth century, concludes, "The scholion on Dem. 24.20 reflects the practice in the Hellenistic period." Why the Dem. 24.20 scholion is regarded as a nugget seems mysterious, since it has been treated by many scholars as only one item among many, and suggests, if not a solution, at least a review of the problem. Indeed, D. M. MacDowell, *Demosthenes Against Meidias* (Oxford 1990) 83, finds the Demosthenes scholia "to be mainly the work of rhetoricians of the third and fourth centuries A.D." with little historical value. It is surprising that such great attention should be given to the "discovery" of such an inferior source, although other scholia are similar.

Solutions in the recent literature have been offered with little regard for the irregularities in the festival or kat' archonta calendar. A state which authorized τετάρτη and ὀγδόη embolimoi days with corresponding exairesimoi ones would hardly authorize its ekklesia to meet according to such an irregular schedule. Meritt (*Hesperia* 32 [1963] 437) charts a year with one month (Elaphebolion) having 42 days and the following one (Mounichion) only 17. In the table (No. 1 of Chapter I) of triple dates, the ekklesia met according to the prytany dates. In No. 6, the festival calendar

decrees according to the prytany dates. His list is particularly valuable because he discusses the various alternative restorations offered.

was retarded by 30 days as late as the eleventh month. We can be sure that there could be no pattern to the schedule of ekklesias if they were required to meet according to such a calendar. Confusion has resulted because the texts of Demosthenes and Aischines give dates of meetings in terms of the festival calendar. See Chapter I p. 34.

The testimonia for the conflicting evidence, largely lexicographical, for the number of meetings and the calendars which regu-lated them, fall into two distinct groups: 1. The Aristotelian one that there were four ekklesias per prytany. 2. There were three ekklesias per month (of the festival calendar); all were called κυρία; in contradistinction to the three κύριαι, there were σύγκλητοι ἐκκλησίαι.

Sources which specify four meetings of the ekklesia in each prytany include:

1. Aristotle *Ath. Pol.* 43.3:

> οἱ δὲ πρυτανεύοντες αὐτῶν πρῶτον μὲν συσσιτοῦσιν ἐν τῇ θόλῳ, λαμβάνοντες ἀργύριον παρὰ τῆς πόλεως, ἔπειτα συνάγουσιν καὶ τὴν βουλὴν καὶ τὸν δῆμον, τὴν μὲν οὖν βουλὴν ὅσαι ἡμέραι, πλὴν ἐάν τις ἀφέσιμος ᾖ, τὸν δὲ δῆμον τετράκις τῆς πρυτανείας ἑκάστης.

In the rest of the chapter, Aristotle outlines the agenda for each of the four ekklesias. In 62.2, he says that the demos was paid 9 obols for the κυρία ekklesia, 6 for the others.

2. Harpokration s.v. κυρία ἐκκλησία (text of J. J. Keaney):

> (K 100) Κυρία ἐκκλησία· Ὑπερείδης ἐν τῷ Κατὰ Δημέου ξενίας (F 89 J), εἰ γνήσιος. τίνες δὲ αἱ κύριαι ἐκκλησίαι Ἀριστοτέλης δεδήλωκεν ἐν τῇ Ἀθηναίων πολιτείᾳ (43.3–4) λέγων: τοὺς πρυτάνεις συνάγειν τὴν βουλὴν καὶ τὸν δῆμον, τὴν μὲν βουλὴν ὁσημέραι, πλὴν ἐάν τις ἀφέσιμος ᾖ, τὸν δὲ δῆμον τετράκις τῆς πρυτανείας ἑκάστης. "προγράφουσι δέ"

> φησί "καὶ κυρίαν ἐκκλησίαν, ἐν ᾗ δεῖ τὰς ἀρχὰς ἀποχειροτονεῖν οἳ δοκοῦσι μὴ καλῶς ἄρχειν, καὶ περὶ φυλακῆς τῆς χώρας. καὶ τὰς εἰσαγγελίας ἐν ταύτῃ τῇ ἡμέρᾳ τοὺς βουλομένους ποιεῖσθαι" φησι, καὶ τὰ ἑξῆς.

Harpokration is dated in the time of Verus in the second century of our era.

3. Pollux 8.95 (ii cent. A.D.): τὸν δὲ δῆμον τετράκις ἑκάστης πρυτανείας. There follows a lengthy account of the agenda of the four ekklesias with some statements not found in Aristotle.

4. Suidas 2760 (x cent. A.D.). As A. Adler notes, taken from Harpokration: τὸν δὲ δῆμον τετράκις τῆς πρυτανείας ἑκάστης.

5. The late Byzantine scholar Maximus Planudes, who wrote scholia on Hermogenes, explains that there were four regular meetings of the ekklesia each prytany, but that if more were needed in a matter of great concern, a synkletos ekklesia was invoked (in *Rhetores Graeci* V p. 509.20 ff., ed. C. Walz):

> ὡρισμέναι ἦσαν ἐκκλησίαι κατὰ πρυτανείαν ἑκάστην τέσσαρες· δέκα δὲ οὐσῶν 'Αθήνησι φυλῶν ἐπρυτάνευον αἱ μὲν πρῶται λαχοῦσαι φυλαὶ τέσσαρες ἀνὰ τριάκοντα ἓξ ἡμέρας, αἱ δὲ λοιπαὶ ἓξ ἀνὰ τριάκοντα πέντε· μετὰ οὖν τὰς ὡρισμένας τοῦ δήμου συνόδους ἐν ταῖς ἐκκλησίαις, εἴ τι ἀναγκαῖον ἐμπέσοι πρᾶγμα, περὶ οὗ τὸν δῆμον ἐχρῆν συνελθόντα γνῶναι, σύγκλητος ἐγένετο ἐκκλησία. προκατακεχρῆσθαι οὖν ἐστιν τὸ ἤδη τὰς ὡρισμένας ἐκκλησίας παρεληλυθέναι.

The synkletos ekklesia was not one of the four.

The focus of Book 8 of Pollux was on the Athenian constitution in which he offers a thesaurus of terms. More importantly, the significant feature of Harpokration is that he cites Philochoros forty-nine times and Didymus forty, as well as Theophrastos *Laws* fifteen. See J. J. Keaney's *Index Locorum*, pp. 277 ff. We can be reasonably certain that these sources did not contain a contradictory record.

Sources which specify three meetings of the ekklesia in each month of the festival calendar:

1. Photius (c. 810 – c. 893) in his *Lexicon* s.v. Κυρία ἐκκλησία does not refer to four meetings, but writes, ἄλλοι δέ φασιν καθ' ἕκαστον μῆνα ἐκκλησίας εἶναι τρεῖς, a statement which he apparently did not endorse. For Photius' sources, see the entry in *OCD*[3] (1996) p. 1175. N. G. Wilson, *Scholars of Byzantium* (1983) 91, writes:

> The ultimate source appears to be the glossary attributed to Cyril, and various additions have been made to it. One of these derives from a lost dictionary known as 'the enlarged *synagoge*', of which reflections are seen in two Paris manuscripts (Coislin 345 and 347). The earlier of the two is written in a hand of archaic appearance that might be regarded as contemporary with Photius himself. By using these sources he was in practice drawing on various Atticist lexica; apart from Diogenianos, mentioned by him in the dedicatory letter, he was using at second hand material from Aelius Dionysius, Pausanias, and Phrynichus. It can be seen how Photius took short series of entries first from one source, then from another, without blending them so as to respect alphabetical order precisely.

2. Scholia to Aristophanes *Ach.* 19 (Text of R. G. Wilson No. 19a):

> **19a.** Κυρίας ἐκκλησίας: ἐν ᾗ ἐκύρουν τὰ ψηφίσματα. εἰσὶ δὲ νόμιμοι ἐκκλησίαι αἱ λεγόμεναι κύριαι τρεῖς τοῦ μηνὸς Ἀθήνησιν. ἡ πρώτη καὶ ἡ δεκάτη καὶ ἡ τριακάς. εἰσὶ δὲ καὶ πρόσκλητοι συναγόμεναι κατά τινα ἐπείγοντα πράγματα. αἱ μὲν οὖν νόμιμοι καὶ ὡρισμέναι ἐκκλησίαι κύριαι λέγονται,

ὡς ἔφαμεν, αἱ δὲ πρὸς τὰ κατεπείγοντα συναγόμεναι σύγκλητοι. REΓ

The oldest of the codices which give this or a similar text is the Ravennas, dated in the tenth century. Rutherford gives the following translation:

> **19. κυρίας ἐκκλησίας**: in which they ratified (ἐκύρουν) decrees. **κυρίας ἐκκλησίας**: there are at Athens three regular meetings of the popular assembly (those called κύριαι) each month, the first, the tenth, and the thirtieth. There are also special meetings, called together on any pressing business. Now the regular and set meetings are called κύριαι, as we have said, whereas those on urgent business are called σύγκλητοι.

3. Scholia to Demosthenes 24 *Against Timokrates* 20 (Text of M. R. Dilts):

> τῇ ἑνδεκάτῃ] κατὰ μῆνα τρεῖς ἐκκλησίας ἐποιοῦντο βουλευόμενοι περὶ τῶν ἐν τῇ πόλει πραγμάτων, πλὴν εἰ μὴ ἄρα ἀνάγκη τις κατέλαβε πολέμου, ὥστε καὶ περὶ ἐκείνου ἄλλην ἐκκλησίαν ποιῆσαι πλέον τῶν ὡρισμένων. καὶ ἐγίνετο ἡ πρώτη ἑνδεκάτῃ τοῦ μηνός, ἡ δὲ δευτέρα περὶ τὴν εἰκάδα, ἡ δὲ τρίτη περὶ τὴν τριακοστήν.

4. Scholion in Demosthenes 18 *Crown* 73 (Text of Dindorf p. 289; not in Dilts):

> ἐν τῇ ἐκκλησίᾳ] σύγκλητος ἐκκλησία λέγεται ἡ γενομένη διά τι ἐξαίφνης κατεπεῖγον· τρεῖς γὰρ ἐκκλησίαι τοῦ μηνὸς γίνονται ὡρισμέναι, ἡ δὲ σύγκλητος οὐχ ὡρισμένη.

Dilts, "Palaeologian Scholia on the Orations of Demosthenes," *Cl. & Med.* 36 (1985) 257–259, explains that he excluded scholia reprinted by Dindorf from the sixteenth-century edition of Morel which were Palaeologian additions to Scholia Demosthenica.

5. Scholion in Demosthenes 19 *False Legation* 123 (Text of Dilts):

> μὴ σύγκλητος ἐκκλησία γένηται] ὅτι τρεῖς ἐκκλησίαι τοῦ μηνὸς γίνονται ὡρισμέναι· ἡ δὲ σύγκλητο· οὐχ ὡρισμένη.

For other references to three ekklesias, see 19.154 (No. 325, Dilts) and 20.94 (No. 209, Dilts).

6. Scholion to Aischines 1 *Timarchos* 60:

> γίνονται δὲ ἐκκλησίαι τρεῖς τοῦ μηνὸς αἱ λεγόμεναι κύριαι, ἃς ἐκ τῶν νόμων ἔχουσιν ἀναγκαίως τελεῖν. ἐπὰν δὲ αἰφνίδιόν τι προσπέσῃ, ἐκκλησιάζουσι μέν, καλεῖται δὲ σύγκλητος, εἰ δ' ἄρα καὶ ἐκκλησία, οὐ κυρία.

7. Scholion to Aischines 3 *Ktesiphon* 24:

> τρεῖς γὰρ ἐκκλησίαι κατὰ μῆνα ἐγένοντο.

8. The Souda, or Souidias, about the end of the 10th century of our era:

> 470 Ἐκκλησία κυρία: οὕτω λέγεται κυρία, ἐν ᾗ ἐκύρουν τὰ ψηφίσματα. εἰσὶ δὲ νόμιμοι ἐκκλησίαι, αἱ λεγόμεναι κύριαι, τρεῖς τοῦ μηνὸς Ἀθήνησιν, ἡ πρώτη καὶ ἡ ι´ καὶ ἡ τριακάς. εἰσὶ δὲ καὶ προσκλητοῖς ἐναγόμεναι κατά τινα ἐπείγοντα πράγματα. αἱ μὲν οὖν νόμιμοι καὶ ὡρισμέναι κύριοι καλοῦνται, αἱ δὲ πρὸς τὸ κατεπεῖγον συναγόμεναι σύγκλητοι.

The scholiasts to Aristophanes *Ach.* and the Souda are in agreement, the 1st, 10th, and 30th.

9. B. B. Rogers, in his introduction to Aristophanes *Ach.* (1910) p. xxvii, writes, "Ulpian (in his Exegesis of Demosthenes against Timocrates 22, p. 706) says that the three Assemblies were held on the eleventh, the twentieth, and the thirtieth of each month."

An 1820 edition by R. Mounteney proved to have only the scholia to the Philippics and Olynthiacs. LSJ^9 p. xxxvii cite Ulpian for an edition by H. Wolf (Basel 1572). The entries for Ulpian in *RE*, *Der kleine Pauly* and *OCD*^{3} (1996) provide no later text. M. R. Dilts, "The Manuscript Tradition of the Scholia Ulpiani on Demosthenis In Timocratem," *TAPA* 105 (1975) 35–50, studies the manuscript tradition, suggesting on p. 47 the need for a new edition of the scholia. However, F. Dübner, in the Adnotatio (pp. 389–390) to his 1877 edition of the scholia of Aristophanes, quotes the text of Ulpian in a note on *Ach.* 18:

> ἡ πρώτη, καὶ ἡ δεκάτη, καὶ ἡ τριακὰς etiam Suidas. At aliter Ulpianus, Demosthenis interpres, in Timocrateam; qui primam ἐκκλησίαν κυρίαν apud Athenienses non primo die mensis, sed undecimo habitam esse tradit. Verba ejus haec sunt: Ἰστέον γὰρ, ὅτι κατὰ μῆνα τρεῖς ἐκκλησίας ἐποιοῦντο, βουλευόμενοι περὶ τῶν ἐν πόλει πραγμάτων. ... καὶ ἐγένετο ἡ πρώτη, ἐνδεκάτῃ τοῦ μηνός. ἡ δὲ δευτέρα, περὶ τὴν εἰκοστήν. ἡ δὲ τρίτη, περὶ τὴν τριακοστήν. Ex hoc igitur loco Ulpiani Petitus De leg. Attic. p. 196 scholiasten nostrum emendandum esse censet: cui assentior.

In his Praefatio to the Teubner *Scholia Demosthenica* 1 (1983), M. R. Dilts observes that many of the scholia to Demosthenes are to be attributed to Ulpianus of Ascalon.[2] J. E. Sandys, *A History of Classical Scholarship* 1^2 (1906) 356, wrote, "Some of the extant scholia on Demosthenes bear the name of Ulpian. They are of little value, and probably belong in part alone to the eminent Sophist, ... who taught rhetoric at Emesa and Antioch, under Constantine." In turn, N. G. Wilson, *OCD*^{3} (1996) 1570, writes, "Ulpianus ... in the reign of Constantine (AD 324–337) ... is the

[2]Cf. Dilts, "Palaeologian Scholia on the Orations of Demosthenes," *Cl. et Med.* 36 (1985) 257–259.

reputed author of scholia to eighteen speeches of Demosthenes; they are of little independent value."

With testimonia about the ekklesias in the festival month, but not about the number, the following passage from Harpokration, who says that the ekklesias were "habitual and by month" with the exception of the synkletos, may be included (text of J. J. Keaney):

> (Σ 59) Σύγκλητος ἐκκλησία· τῶν ἐκκλησιῶν αἱ μὲν ἐξ ἔθους καὶ κατὰ μῆνα ἐγίνοντο· εἰ δέ τι ἐξαίφνης κατεπείξειεν ὥστε γενέσθαι ἐκκλησίαν, αὕτη ἐκαλεῖτο σύγκλητος ἐκκλησία· Δημοσθένης ἐν τῷ Κατ' Αἰσχίνου (19.122).

In his 1853 edition, Dindorf suggests that Harpokration found this statement as a gloss on the spurious decree in the text of Demosthenes 18.37.

In the first collection, which might be called Aristotelian, we have uniformity for four ekklesias in each prytany with one being κυρία. The second collection displays almost equal uniformity: three ekklesias in each festival month. All were called κύριαι according to the Souda and two other sources. These three were held on the 11th, around the 20th, and around the 30th in two sources; on the 1st, 10th, and 30th in two others. In addition, a special ekklesia called synkletos might be called.

The first group can be traced to Aristotle *Ath. Pol.*; the second group cannot be traced to any source earlier than the rhetoricians of the third and fourth centuries of our era. We are not dealing with authorities who walked about the akropolis and agora and examined the prescripts of Athenian decrees.

EKKLESIAS IN PRYTANY VIII OF 347/6

A cornerstone of Mogens H. Hansen's argument about the number of ekklesias, or what he calls a "precious piece of information" (*The Athenian Assembly* [Oxford 1987] 21), relates to a passage in Demosthenes concerning the departure of the second embassy to Philip in Mounichion (X) 347/6 B.C. In 19 *de fal. leg.* 154, Demosthenes relates that the Athenians had used up the number of ekklesias (ἐπειδὴ γὰρ ἐκκλησία μὲν οὐκέτ' ἦν ὑπόλοιπος οὐδεμία διὰ τὸ προκατακεχρῆσθαι) and that he had procured a decree of the boule which was empowered to act until the next ekklesia, directing the embassy to depart at once and ordering Proxenos to convey the embassy to Philip. Such empowerment is well attested epipgraphically: τὴν βουλὴν κυρίαν εἶναι ψηφίζεσθαι ὅτι ἂν αὐτῆι δοκῆι ἄριστον εἶναι. See *IG* II^2 127.34 (356/5); 204.85 (352/1); 438.8 (post 336/5). Cf. II^2 43.34. For the verb προκατακεχρῆσθαι applied to ekklesias, see the passage of Maximus Planudes quoted above (p. 187). A. Schaefer, *Demosthenes und seine Zeit*[2] II (1886) 248, devotes a paragraph to this psephisma of the boule which was passed on Mounichion (X) 3 346 B.C. (Aischines 2 *Embassy* 91–92).

Whereas Demosthenes 19.154 only says that the ambassadors before Mounichion 3 were wasting time in Athens, Aischines 2.90 explains the delay of this second embassy to Philip as προξενίας κατασκευαζόμενοι. Grote (IX.403 n. 2 in Second London edition, chap. 89) remarks that this explanation "is altogether without value".[3] Grote writes of this delay (IX.403):

> Impressed with this necessity for an immediate interview with Philip, Demosthenes urged his colleagues to set out at once. But they resisted his remonstrances, and chose to remain at Athens; which, we may remark, was probably in a state of rejoicing and festivity in consequence of the recent peace. So reckless was their procrastination and reluctance to depart, that on the 3d of the month Munychion (April—nine days after the solemnity of oath-taking before Antipater and Parmenio) Demosthenes made com-

[3]Demosthenes 18 *Crown* 82 charges that in 343/2 Aischines entertained the envoys of the tyrants of Oreos and Eretria as their πρόξενος. See Schaefer II^2 249.

plaint and moved a resolution in the Senate, peremptorily ordering them to begin their journey forthwith.

The ekklesia clearly was not meeting according to the festival calendar, for any allotted meetings could not have been used up by the third of Mounichion. Ekklesiasts drew pay for attendance; so there must have been a budget which could easily have been exhausted by the meetings in Elaphebolion. The ekklesias on the 18th and 19th for the Peace of Philokrates must have been widely attended. The meetings began at dawn,[4] and citizens in Acharnai, Marathon, and Sounion must have left home at a very early hour. The two orators make no distinction about the nature of the ekklesias, kyria, synkletos, etc., except to say that the principal meetings were in the theater of Dionysos.

From the pleadings of the two rival orators some years afterwards, with glaring contradictions between the two as well as with themselves, the calendar of the month in which the Peace of Philokrates was passed has been worked out by three authorities: G. Grote, Chapter 89 = XI.385 ff. of the Second London edition; A. Schaefer, *Demosthenes und seine Zeit*[2] II (1886) 248; and W. W. Goodwin, *Demosthenes on the Crown* (Cambridge 1901) 242 ff. The three scholars are in accord except about numerals to be read in a passage in Aischines.[5] The ekklesia met on at least five

[4]For a collection of passages listing dawn as the time of the ekklesia, se R. G. Ussher on Aristophanes *Ekklesiazousai* lines 19–21.

[5]Grote (XI.396 n. 1 in the Second London edition), our closest student of the literary testimonia, in a note on Elaphebolion 25, suggested according to unemended texts that there were two ekklesias on successive days, i.e. by the correct accounting of days, on 25 and 26: "This date is preserved by Aeschines adv. Ktesiph. p. 64 c. 27. ἕκτῃ φθίνοντος τοῦ 'Ελαφηβολιῶνος μηνὸς, etc. In the earlier oration (De Fals. Leg. p. 40 c. 29) Aeschines states that Demosthenes was among the Proedri or presiding senators of a public assembly held ἑβδόμῃ φθίνοντος—the day before. It is possible that there might have been two public assemblies held, on two successive days (the 23d and 24th, or the 24th and 25th, according as the month Ela-

days in the month of Elaphebolion, in what I shall argue was the eighth prytany, and quite possibly on one or two additional days.

Envoys returned from the court of Philip about the first of Elaphebolion (IX) 346 B.C. They brought a letter from Philip expressing great friendship, and received a complimentary vote of the boule and ekklesia with an invitation to dinner in the prytaneion on the motion of Demosthenes. A decree was enacted that safe-conduct should be granted to Philip's envoys and herald, who were on their way to Athens, and that the prytaneis should call a special meeting of the ekklesia to be held on the eighth of Elaphebolion (IX) to hear the envoys from Makedonia and to discuss terms of peace. Since the Makedonian envoys, Antipater, Parmenion, and presumably Eurylochos, did not arrive on Elaphebolion (IX) 8,[6] Demosthenes proposed and carried a fresh

---

phebolion happened in that year to have 30 days or 29 days), and that Demosthenes may have been among the Proedri in both. But the transaction described (in the oration against Ktesiphon as having happened on the latter of the two days)—must have preceded that which is mentioned (in the Oration De Fals. Leg.) as having happened on the earlier of the two days; or at least cannot have followed it; so that there seems to be an inaccuracy in one or in the other. If the word ἕκτη in the oration against Ktesiphon, and ἑβδόμη in the speech on the False Legation, are both correct, the transactions mentioned in the one cannot be reconciled chronologically with those narrated in the other. Various conjectural alterations have been proposed. See ..." The correct references to Aischines are 3.73 and 2.90. See also below.

[6]The date of Elaphebolion 8 is to be explained by the fact that the Dionysia began on the 10th and two days were apparently required for ekklesias deliberating on peace. Cf. Thucydides 1.44.1 for two successive ekklesias in the negotiations with Korinth, and R. B. Richardson's note on Aischines 3.68. Aischines (3 *Ktesiphon* 66–67) criticizes Demosthenes for the date when there was a sacrifice to Asklepios and the proagon ὃ πρότερον οὐδεὶς μέμνηται γεγονός. The Asklepieion was on the south side of the akropolis, a short distance removed from the theater, and its devotees doubtless performed a sacrifice. According to the scholiast, the proagon was for the Great Dionysia and held in the small Odeion, adjacent to the theater. "Each poet mounted a temporary platform with his actors and announced the subject of the plays which he was about to produce:" A.

decree in this ekklesia, fixing two days (Elaphebolion 18 and 19) after the Dionysiac festival for ekklesias in the theater of Dionysos. Grote (XI.390) has the following note on these meetings:

> But it seems to have been a common practice with the Athenians, about peace or other special and important matters, to convene two assemblies on two days immediately succeeding: all that the synod here recommended was, that the Athenians should follow the usual custom—προγράψαι τοὺς πρυτάνεις ἐκκλησίας δύο κατὰ τοὺς νόμους, etc. That two assemblies, neither less nor more should be convened for the purpose, was a point of no material importance except that it indicated a determination to decide the question at once—*sans desemparer.*

The modern reference to the Greek passage is Aischines 2 *Embassy* 60, correcting the plural to the singular (κατὰ τὸν νόμον).[7] Demosthenes personally escorted the envoys to the theater, where curtains had been provided to shield them from the early morning air and cushions to cover the stone seats. And when they departed for home, he hired three yokes of mules for them and escorted them on horseback to Thebes. A few days after the peace was voted, in an ekklesia held on Elaphebolion 25 in which Aischines says Demosthenes presided, it was voted that the representatives of the allies of Athens then present should take the oath of ratifica-

---

Pickard-Cambridge, *Dramatic Festivals of Athens*[2] (1968) 67. Tomlinson, *OCD*[3] 1060, writes, "There are no traces of the provision for the audience, but Plutarch says it contained seats." There were earlier examples of ekklesias on so-called festival days, some cited by Pickard-Cambridge, to prove that Aischines' statement was a highly rhetorical exaggeration. In referring to 'festival' days, it is important to make a distinction between those with a procession and games and participated in by a large element of the populace and those involving a sacrifice if with a limited audience. The proagon was not a 'festival', but for limited attendance.

[7]The full text reads: προγράψαι τοὺς πρυτάνεις ἐκκλησίας δύο κατὰ τὸν νόμον ἐν δὲ ταύταις βουλεύσασθαι περὶ τῆς εἰρήνης 'Αθηναίους. Deliberations concerning peace by law required two (δύο) ekklesias.

tion. Grote (11.396 n. 1 = Chap. 89), as we have seen above, suggests that there were two ekklesias on successive days. See also below on Aischines 2.90. Whether there was an ekklesia between the 19th and the 25th to receive the envoy of Kersobleptes, as Schaefer (II^2.242) seems to suggest,[8] is not specifically attested. The Phokians gave notice either at or after the ekklesia of the 25th that Philip refused to admit the Phokians. After the 25th, Antipater administered the oaths to the allies of Athens in the στρατήγιον. On Mounichion 3, Demosthenes says they had run out of ekklesias.

If 347/6 was an ordinary year, Elaphebolion (IX) 1 would fall on the 237th day of the year (4 x 30 + 4 x 29 + 1), corresponding to Prytany VII.23 (4 x 36 +2 x 35 + 23). If intercalary and Posideon (VI) was the intercalary month, Elaphebolion (IX) 1 would fall on the 267th day of the year, corresponding to Prytany VII.35 (4 x 39 + 2 x 38 + 35). Dinsmoor (*Archons of Athens* [1931] 428) claims that 347/6 was intercalary according to his "Metonic" cycle. Incidentally, G. T. Toomer, in *OCD*[3] (1996) s.v. Meton, is a recent scholar who cautions that Meton's calendar was astronomical and had nothing to do with the festival calendar. If an intercalary year and Posideon was the intercalary month, Mounichion (X) 3 (5 x 30 + 5 x 29 + 3), the 298th day of the year and the date of the psephisma of the boule, would fall on Prytany VIII.28 (4 x 39 + 3 x 38 + 28). If an ordinary year, Mounichion 3 would fall on the 269th day (5 x 30 + 4 x 29 + 3), corresponding to Prytany VIII.20 (4 x 36 + 3 x 35 + 20). If in an ordinary year, Prytany VIII.1 would fall on Elaphebolion (IX)

[8]Cf. Goodwin p. 257. In 2.58, Aischines writes, ἀλλ' 'Αθήνησι· ταῖς δὲ ξενικαῖς πρεσβείαις ἡ βουλὴ τὰς εἰς τὸν δῆμον προσόδους προβουλεύει· (Loeb: "Now for embassies from foreign states an opportunity to address the assembly of the people is always provided by a decree of the senate.").

13±; if intercalary, on Elaphebolion 5±. Thus, on Mounichion (X) 3, Prytany VIII, if in an ordinary year, still had 15 days to run; if in an intercalary year, 7. These calculations are based on the assumption that Aristotle's description of the regular length of prytanies (*Ath. Pol.* 43.2) in an ordinary year is correct, and that such regularity extended to intercalary years. I also assume that there was no tampering with the lunar festival calendar.

For the eighth prytany, held by the tribe Aigeis, of the year 347/6, the archonship of Themistokles, which saw the passage of the Peace of Philokrates, we have two epigraphical decrees. One, *IG* II² 213 (= *SEG* 21.263; 34.259; 37.102) was the renewal of an alliance with Mytilene as a result of envoys who came from Phaidros, renewing friendship with Athens as a result of the restoration of democracy. The day of the prytany is not given.

In a decree dated in the eighth prytany of Aigeis of 347/6, honors were voted to Spartolaos and Pairisades, joint kings of Bosporos and their brother Appollonios: *IG* II² 212.[9] In lines 53–59, we learn that a debt was owed by the Athenians to the Bosporans, and the proedroi were instructed to put the matter on the agenda of the ekklesia on Elaphebolion 18. In the many pages of his collection of what he calls the "Audienzformel," W. Larfeld, *Handbuch der griechischen Epigraphik* 2.2 (1902) 667 ff., this example is the only one specifying a date in the festival calendar, in contrast with the usual εἰς τὴν ἐπιοῦσαν (or πρώτην) ἐκκλησίαν phrase. No subsequent example has turned up. Lewis (*BSA* 50 [1955] 25) has argued accordingly that there must have been one (or more) ekklesias intervening between that of the passage of *IG* II² 212 and that on Elaphebolion 18. It was only on Elaphebolion 8 that ekklesias were fixed for the 18th and 19th. It follows that since the meeting on the 8th is in Prytany VIII (Aigeis), the year

[9]For these Bosporan kings, see Gardiner-Gordon, *Historia* 35 (1986) 207–209.

was intercalary.[10] The intervening ekklesia must be the statutory one held annually on the day after the Pandia, which is dated by Lewis between the 15th and 17th and by D. M. MacDowell, *Demosthenes Against Meidias* (Oxford 1990) 227–228, on the 17th. Thus, we have ekklesias in Prytany VIII on the 8th, 15–17th, 18th, 19th, 25th, and possibly a later one.

With respect to these affirmed ekklesias, none give any clue as to when such subjects as catalogued in *Ath. Pol.* 43.4–6 for the regular agenda of each prytany were treated: the grain supply (a subject of great importance), claims about inheritances, ostracism (final vote in the eighth prytany: Philochoros 328 frg. 30 with Jacoby's note), petitions, etc. We have a host of examples of the name of Themistokles, archon in 347/6, in epigraphical contexts, including: Treasurers of Athena (*IG* II^2 1441 line 4; 1443.92,

[10]Lewis, *BSA* 50 [1955] 25, earlier deduced that 347/6 was an intercalary year at Athens, but his reasoning is faulty. He argues that 345/4 was ordinary because R. Vallois has determined that a secretary of Athenian shipbuilders at Delos was paid 355 drachmai at the rate of a drachma per day for his year's work. Moreover, he says that the restored date for an ekklesia in this same year can be calculated to require an ordinary year; Schweigert, *Hesperia* 8 (1939) 172–173. The date is entirely restoration, but accords with the stoichedon order in this small fragment. For the year 346/5, Lewis notes that Dinsmoor, *Archons of Athens* (1931) 355, had computed from a calendar equation that the year at Samos was ordinary and assumes that the Samians were using the calendar of the home city. The inscription in question is an inventory of the temple of Hera, dated to the Samian archon: Michel no. 832 lines 57–58. Lewis infers from the Delian and Samian evidence that, since 346/5 and 345/4 were ordinary in one or the other of these two dates, they were also ordinary at Athens, and that it was likely that an intercalary year would precede two ordinary ones. If A. E. Samuel (*Greek and Roman Chronology* [1972]) is correct, this line of reasoning fails. He writes of Delos (p. 99), "Delos ... maintained until 166 a separate calendar with a number of months different in name from those of Athens. The calendar year began after the winter solstice," etc. Of Samos, he notes (pp. 120–121) that this island shared the names of only four months with Athens.

104, 206); Treasurers of Athena and Other Gods (*IG* II^2 1455.17–18; *SEG* 15.120.26); Commissioners of Brauronian sanctuary (*IG* II^2 1514, 1519, 1521); Naval Commissioners (*IG* II^2 1622.446–447); Dionysia (*IG* II^2 1318).[11]

According to the codices of Aischines 2 *Embassy* 90, Chares, the Athenian commander in the Hellespont,[12] sent a letter saying that Philip had taken Hieron Oros in the Thrakian Chersonese and Kersobleptos had lost. The date of the capture is given as ἕκτῃ φθίνοντος (= IX 25). Aischines follows this statement by referring to Demosthenes as presiding over the ekklesia on the 24th (ἑβδόμῃ φθίνοντος). As noted above (n. 5), Grote followed the codices and suggested that there were ekklesias on successive days. However, since we learn from 3.73 that there was an ekklesia on IX 25, where the date is twice given, Spengel (*RM* 2 [1828] 380) and Schaefer (II^2 245) and recent editors reverse the numerals: ἑβδόμῃ ... ἕκτῃ. The second numeral may well be emended, but if the capture of Mount Athos took place on the 24th, as the emended text has it, news of the capture could hardly have reached Athens by sea for the ekklesia the next morning. If later, an additional meeting of the ekklesia may have been required, and the text of Aischines certainly implies that there was an ekklesia addressing the capture of Hieron Oros. If so, we have not five, but six ekklesias in Prytany VIII.

At the time of Demosthenes' statement about the ekklesias being used up, the demos, but not Demosthenes, had faith in the Peace of Philokrates and hence did not anticipate that Philip would advance on Thermopylai and destroy Phokis, so saw no need of another ekklesia. The most that Demosthenes was able to

[11]Cf. A. G. Woodhead, *The Athenian Agora* XVI (1997) 106–107.

[12]Chares is not mentioned again by Aischines, and not at all by Demosthenes in this context. For his generalship in this year, see Demosthenes 9 *Third Phil.* 5; Polyainos 3.13.2; Schaefer II^2 178 n. 3, 179.2, 246.3.

procure was a decree of the boule sanctioning the departure of the second embassy, which ultimately found Philip at Pella with the humiliating result; see, for example, W. W. Goodwin, *Demosthenes De Corona* 259 ff. Demosthenes' statement is a rhetorical explanation.

Parenthetically, I note that since the year of Themistokles (347/6) was intercalary, the Athenians were not observing a rule that the Athenian year began with the first new moon after the summer solstice. R. A. Parker and W. H. Dubberstein, *Babylonian Chronology* (1942) 33, give the following sequence of dates at Babylon for 347/6, assuming that the year began after the summer solstice:

> 6/8 // 7/8, 8/6, 9/5, 10/5, 11/3, 12/3, 1/2, 1/31, 3/2, 3/31, 4/29, 5/29 // 6.27

Allowing for the differences in time between Athens and Babylon, the year 347/6 would have had only 12 months, if they followed the rule Plato (*Laws* 767C) laid down for his ideal state. See also the New Moon table for 347/6 in E. J. Bickerman, *Chronology of the Ancient World* (London 1968) p. 119. Oddly, the tables were omitted in the second edition.

DEMOSTHENES 24.25 AND EKKLESIAS IN 353 B.C. AND EARLIER

In the *Oration against Timokrates*, Demosthenes, in referring to old established laws (πολὺν χρόνον), states (24.25):

> καὶ πρῶτον μὲν ἐφ' ὑμῖν ἐποίησαν διαχειροτονίαν, πότερον εἰσοιστέος ἐστὶ νόμος καινὸς ἢ δοκοῦσιν ἀρκεῖν οἱ κείμενοι· μετὰ ταῦτα δ' ἂν χειροτονήσητε εἰσφέρειν, οὐκ εὐθὺς τιθέναι προσέταξαν, ἀλλὰ τὴν τρίτην ἀπέδειξαν ἐκκλησίαν, καὶ οὐδ' ἐν ταύτῃ τιθέναι δεδώκασιν, ...

In the document inserted into the previous paragraph, we have this text (24.21):

> ἐὰν δέ τινες τῶν νόμων τῶν κειμένων ἀποχειροτονηθῶσι, τοὺς πρυτάνεις, ἐφ' ὧν ἂν ἡ ἐπιχειροτονία γένηται, ποιεῖν περὶ τῶν ἀποχειροτονηθέντων τὴν τελευταίαν τῶν τριῶν ἐκκλησιῶν· τοὺς δὲ προέδρους, οἳ ἂν τύχωσι προεδρεύοντες ἐν ταύτῃ τῇ ἐκκλησίᾳ, χρηματίζειν ἐπάναγκες πρῶτον μετὰ τὰ ἱερὰ περὶ τῶν νομοθετῶν, ...

One notes that for those who accept the validity of the latter passage, the reference to officials who held office for the length of a prytany would prove that the number of ekklesias would be according to the prytany calendar, not the festival calendar of lunar months.

Demosthenes' phraseology, μετὰ ταῦτα ... ἀλλὰ τὴν τρίτην ... ἐκκλησίαν and the τὴν τελευταίαν τῶν τριῶν ἐκκλησιῶν of the inserted document have resulted in a monumental debate which has persisted for over a hundred years. The focus is on two problems, the validity of the inserted document, and whether Demosthenes is referring to three or four ekklesias per prytany, which involves the problem of the inclusive or exclusive count with the ordinal τὴν τρίτην.

First, the validity of the inserted document. Hansen and Mitchel (*SO* 59 [1984] 15 and 18) write that Demosthenes in oration 24.25 is "paraphrasing the law" found in the document inserted into 24.21. There is a basic misunderstanding here. The reverse is the case.

In the text of 24 *Timokrates*, sixteen documents, chiefly laws, have been inserted. In some of the orations of Demosthenes, documents are inserted into the text (e.g. 18 *Crown*; 21 *Meidias*; 24 *Timokrates*). In others, they are omitted (e.g. 19 *Embassy*; 20 *Leptines*), and we have only lemmata for nomoi, psephisma, and the like. In *Against Aristokrates*, the insertions were not made after 23.87. In the case of the three orations of Aischines, citations of psephismata and nomoi are contained in the MSS. of only the speech *Against Timarchos*, and these are rejected by C. D. Adams in the introduction to his Loeb volume as spurious. The MSS. of Lysias do not contain citations.

The document in 24.20–23 is referred to by Demosthenes as νόμοι, and they clearly are not in chronological order. In 1954, the Budé editors Navarre/Orsini, provided a note on this passage:

> 1. Comme l'indique l'orateur au § 19 *fin.*, il s'agit de plusieurs textes législatifs se complétant l'un l'autre. L'authenticité de ces documents est généralement admise. Sans doute certaines des dispositions mentionnées ont pu être tirées du contexte démosthénien (par ex. le § 23); d'autres, auxquelles le plaidoyer ne fait qu'une allusion assez vague (§ 30: époque à laquelle l'action législative est admise), y sont énoncées avec plus de précision. Il en est cependant dont l'orateur ne dit mot, mais qui ne semblent pas, malgré tout, avoir pu être inventées de toutes pièces par un faussaire (§ 20, 21 et 22: procédure à suivre pour la revision des textes législatifs et sanctions prévues contre les magistrats convaincus d'avoir sur ce point violé la loi). La dernière prescription (§ 23 *fin.*: élection de cinq σύνδικοι) est confirmée par le *c. Leptine* 146 et *infra* 36. Certaines formules ont été abrégées par le copiste: ainsi ὁ βουλόμενος (23) est une abréviation de la formule officielle: ὁ β– οἷς ἔξεστιν (cf. *infra* 63, 105, textes de loi).

The obviously apocryphal documents in the *De Corona* have received special studies, for which D. F. Jackson and G. O. Rowe in *Lustrum* 14 (1971) 76, give the following summary, with references by numbers to their bibliographical tables:

> Since the first half of the nineteenth century scholars, with few exceptions, have regarded the documents inserted in the oration *On the Crown* (vxiii [sic]) as apocryphal. Yet in the twentieth century interest in these inserts has remained strong with efforts being directed toward determining their date, sources, and authorship. It is now almost certain that the documents first appeared in editions of Demosthenes around the end of the second century B.C. or the beginning of the first and not, as was believed earlier, in the second

> century A.D. The most exhaustive study on the subject, that by Schlaepfer (**336**), has shown that most of the documents are outright fabrications of a forger's imagination or paraphrases of Demosthenes' statements in the text proper. Some few documents, however, would indicate that the forger has derived their content and style from actual decrees. In language and style they are somewhat uniform—a fact which leads Schmid (**335**) to conclude that they are all from the same hand. Treves (**338**), however, and Schlaepfer (**336**) believe that the uniformity is only superficial and thus conclude that the inserts stem from several hands, who, according to Schlaepfer, worked concurrently on their fabrications. The place where the documents were composed is still not agreed upon. Schmid and Treves, who find their legal jargon similar to the language employed by oriental monarchies and their style on the whole Asian rather than Attic, would favor Hellenistic Asia Minor. Schlaepfer, on the other hand, argues that the place of origin is most probably Athens, where two rhetoricians especially devoted to Demosthenes, Pammenes and Menedemos, maintained a school.

For other comments on the fabricated documents, see J. A. Goldstein, *The Letters of Demosthenes* (New York 1968) 17, who notes that a papyrus has texts for documents mentioned only by title in the codices. See also D. M. MacDowell's section on "Documents" in *Demosthenes Against Meidias* (Oxford 1990) 43–47. Sometimes, the forgers deduced their contexts from the speeches, as may have been the case with 24.21.

I pass in review the opinion of several scholars that the passage in 24.25 refers to four ekklesias per prytany, as Aristotle says was the case a quarter of a century later.

Adam Reusch, *De diebus contionum ordinariarum apud Athenienses* (= *Dissertationes Philologicae Argentoratenses* Vol. 3) 1880. 53, wrote:

Quod igitur ad tempora X tribuum attinet in Aristotelis de quattuor contionibus testimonio tuto acquiescere licet. Idem autem confirmari videteur lege in Demosthenis Timocratea servata § 21: τοὺς πρυτάνεις, ἐφ' ὧν ἂν ἡ ἐπιχειροτονία γένηται, ποιεῖν περὶ τῶν ἀποχειροτονηθέντων τὴν τελευταίαν τῶν τριῶν ἐκκλησιῶν collatis oratoris verbis § 25: καὶ πρῶτον μὲν ἐφ' ὑμῖν ἐποίησαν διαχεροτονίαν, πότερον εἰσοιστέος ἐστὶ νόμος καινὸς ἢ δοκοῦσιν ἀρκεῖν οἱ κείμενοι. μετὰ ταῦτα δ' ἂν χειροτονήσητε εἰσφέρειν, οὐκ εὐθὺς τιθέναι προσέταξαν, ἀλλὰ τὴν τρίτην ἀπέδειξαν ἐκκλησίαν, καὶ οὐδ' ἐν ταύτῃ τιθέναι δεδώκασιν, ἀλλὰ σκέψασθαι καθ' ὅ τι τοὺς νομιθέτας καθιεῖτε. Quamquam Westermann qui legem illam omnino fictam esse arbitratur, ex his quoque verbis; τὴν τελευταίαν τῶν τριῶν ἐκκλησιῶν, quam nulla fides sit grammatico qui legem composuerit, adparere statuit. Demosthenem enim verbis τὴν τρίτην ἀπέδειξαν ἐκκλησίαν significasse putat non tertiam contionem post primam, sed tertiam anni contionem prima comprehensa idemque intellexisse etiam legis inventorem, sed ex suae aetatis exemplo pro tertia ultimam posuisse trium contionum per prytaniam habendarum quot habitae essent XII tribuum tempore. Verum id neque in Demosthenem valet neque in legis auctorem quisquis fuit. Nam si bene respicis oratoris verba μετὰ ταῦτα δὲ (i. e. post διαχειροτονίαν in prima contione habendam)—τὴν τρίτην ἀπέδειξαν ἐκκλησίαν, quin Demosthenes ipse significarit tertiam contionem post primam illam legitimam vix ac ne vix quidem dubitari potest. Notum autem est ex Dem. Lept. § 94 (τῷ γραμματεῖ παραδοῦναι, τοῦτον δ' ἐν ταῖς ἐκκλησίαις ἀναγιγνώσκειν) novas leges in compluribus contionibus, minimum duabus, recitari oportuisse, ut unusquisque cum videret, quantus esset numerus legum ferendarum, de nomothetis instituendis consilium recte iniret. In ea autem contione quae ἡ τελευταία τῶν τριῶν vocatur, secundum legem illam recitari leges novae non iam potuerunt, quia ex eadem lege πρῶτον μετὰ τὰ ἱερὰ agendum est de nomothetis instituendis. Itaque neque Demosthenes neque is qui legem composuisse dicitur intellexit tertiam anni contionem, sed

> tertiam post primam, ultimam ex tribus post primam reliquis, i. e. quartam contionem et anni et prytaniae primae.

R. Schoell, *SBAW* 1886.101, supported the position of Reusch:

> Er (Westermann) hält die 'dritte' Ekklesia bei Demosthenes für die dritte der Prytanie: der Ausdruck 'die letzte der drei Versammlungen' soll verrathen, dass der Verfasser der Einlage nur drei ordentliche Gemeindeversammlungen der Prytanie kannte, also von der Voraussetzung der späteren Zeit ausging, da die Bürgerschaft in zwölf Phylen gegliedert war. Indess ἡ τρίτη ἐκκλησία ist im Munde des Redners die drittnächste von dem vorher angedeuteten Zeitpunkt an, wie in der bekannten Formel des Probuleuma ἡ πρώτη ἐκκλησία nicht die 'erste' schlechthin, sondern die 'nächstfolgende' (ἡ ἐπιοῦσα heisst es seit Anfang des 3. Jahrhunderts): als 'die letzte er drei (noch übrigen) Versammlungen' wird verständlich genug die vierte und letzte ordentliche Versammlung der Prytanie bezeichnet.

Reusch refers to the passage in 20 *Leptines* 94 (τοῦτον δ' ἐν ταῖς ἐκκλησίαις ἀναγιγνώσκειν ἵν' ἕκαστος ὑμῶν ἀκούσας πολλάκις ...),[13] for which J. E. Sandys, in his 1890 edition, gives the following commentary:

> **ἐν ταῖς ἐκκλησίαις**] It was at the first meeting of the assembly, held on the eleventh day of the first prytany in the year, that the people decided whether there should be any revision of the laws. At the next two meetings the proposed law had to be publicly recited, and at a third meeting a vote was taken on the question of referring it to a legislative committee. 24 § 25 οὐκ εὐθὺς τιθέναι προσέταξαν, ἀλλὰ τὴν τρίτην ἀπέδειξαν ἐκκλησίαν. (Cf. Schöll, *über attische Gesetzgebung*, p. 101.)

The plural form indicates that there was more than one ekklesia between the first and fourth, as is also suggested by the adverb

[13]Dobree would emend ταῖς to τρισῖν. The ταῖς was omitted in S, the earliest codex. Cobet believes the archetype read ἐν γ′ ἐκκλησίαις.

πολλάκις. The date of the oration is 355 B.C., only two years removed from the date assigned to Demosthenes' reference in 24.25.

In his review of the laws in the inserted document in *JHS* 95 (1975) 66–69, D. M. MacDowell writes, "'The last of the three' perhaps means the third meeting after the one mentioned earlier," with a note that some regard the count as inclusive.[14] The same difference is found in translations. The Budé editors translate τὴν τρίτην ἐκκλησίαν as 'la troisième assemblée', whereas the Loeb editor renders the phrase as 'the next assembly but one'. M. H. Hansen in several publications asserts that only the latter translation is correct. For example, Hansen and F. Mitchel in *SO* 59 (1984) 14–15, write:

> There is no indication whatsoever in any source warranting the inference that ἡ δευτέρα or ἡ τρίτη ἐκκλησία could mean the second or third meeting AFTER this meeting, i.e. the third or the fourth meeting. On the contrary, the Athenians preferred inclusive counting, and ἡ τρίτη ἡμέρα means invariably the day after tomorrow and never the third day from tomorrow, i.e. the fourth day from today.

In Xenophon *Anab.* 3.4.31, τετάρτῃ δ' ἡμέρᾳ is positively 'on the fourth day', as the context proves. In a series of years in Lysias 7.10, τρίτῳ δὲ ἔτει is positively the 'third'. In Xenophon *Hell.* 1.2.6, we have, ἑβδόμῃ καὶ δεκάτῃ ἡμέρᾳ μετὰ τὴν εἰσβολὴν. If δευτέρᾳ ἡμέρᾳ were always the next day, we would hardly find such phrases as τῇ πρώτῃ ἡμέρᾳ μετὰ τὰς ἀντιδόσεις as in Demosthenes [42] 10. Translators render μετὰ δὲ ἡμέρῃ τρίτῃ ἢ τετάρτῃ in Herodotos 1.30 as 'on the third day'.

[14]In his latest pronouncement, MacDowell, in *OCD*³ (1996) 1048, citing himself, Rhodes, and Hansen (the reference to Rhodes is incorrect), says that the information in 20 and 24 is "incomplete and tendentious."

Thucydides (1.18.2) says δεκάτῳ ... ἔτει μετ' αὐτήν (= τὴν μάχην), where the interval is generally considered ten years to the month. Ordinal numerals are frequently found with μετά, ὕστερον, and πρότερον, and the readers will find that translators differ. Students of the Athenian archon-table will know that the phrase δεκάτῳ ὕστερον ἔτει in Plutarch *Mor.* 847E for the archon Pytharatos once created a difference of opinion in the count.[15] ἐπὶ τῷ τρίτῳ (σημείῳ) is the third in a catalogue in Xenophon *Anab.* 2.2.4. I know of no ancient or modern authority on the subject, and Hansen cites none for his 'invariable' rule. In turning to the examples of τρίτος in Herodotos, I note that for Enoch Powell, who has given us our most accurate English translation of Herodotos, ἐς τ. ἡμέρην is 'day after tomorrow' (5.49.9), διὰ τ. ἔτεος, ἡμέρης is 'every other year, day' (2.4.1, 2.37.2, 3.97.2, 4.180.6). τριταῖος is 'on the next day but one'. But in a chronological series in contrast with the dates, τ. is the 'third' for Powell: 1.30.3, 3.131.2, 4.42, 5.42, 6.40.1, 7.171.1, 8.15.1, 9.37.3. τὸ τρίτον is 'for the third time'. I suggest that if the writer is visualizing a list or sequence, as I believe is the case in our Demosthenes passage, he uses exclusive count (πρώτη, δευτέρᾳ, etc.), as the modern does. I note after looking up a number of examples of ordinals with ἔτος, μήν, and ἡμέρα that in examples where translators use inclusive count, the article is almost always omitted, but otherwise with ordinals in a series. Cf. John Thompson, "On the Use of the Article with Ordinal Numerals in Greek," *CR* 20 (1906) 304. Thompson publishes examples in Thucydides. All the ordinals with article, I believe, some in a prepositional phrase, yield exclusive count. One may find exceptions, but Hansen's 'invariable' rule is not universal. I follow Reusch, Sandys, MacDowell in interpreting τὴν τρίτην ἐκκλησίαν

[15]Cf. A. E. Samuel, *Greek and Roman Chronology* (Munich 1972) 211. For exclusive count, see p. 205.

καὶ οὐδ' ἐν ταύτῃ as the third ekklesia after the first, i.e. four ekklesias per prytany.[16]

To judge from the examples in *LSJ*, τετραετής, τετραετηρία were used for duration of time, "of four years", whereas πεντετηρίς as an interval of time by inclusive count was used for a quadrennial festival. However, *LSJ* cites τετραετηρίς in *CIG* 2741.22 and 2812, as well as πετροετηρίς (Thess.) = τετραετηρίς, in *Rev. Phil.* 35. 123 as a quadrennial festival. The confusion was noted by Heath, *Aristarchus of Samos* p. 286. The octaëteris was an eight-year cycle, a period of eight years. The enneadekaeteris was of 19 years, not 18. If one thinks of a period, the count is exclusive. In *IG* II² 333.c.17, ἑκάστου τοῦ ἐνιαυτοῦ ἐν τῆι τετραετίαι is "in the tetraëteris", not "in the 3-year period", as Schwenk, *Athens in the Age of Alexander* 123, takes it. Censorinus (18.3) uses tetraeteris for a four-year period. For the problem of a restoration of διὰ πεν[τ---] in an inscription from Kos, see J. and L. Robert, *REG* 65 (1952) pp. 44–45. For examples with μετὰ, see Plato *Republic* 461D: μετ' ἐκείνην δεκάτῳ μηνὶ καὶ ἑβδόμῳ; Thucydides 3.116.2: πεντηκοστῷ ἔτει ... μετὰ τὸ πρότερον ῥεῦμα. See also Gomme's note on Thucydides 1.18.2: δεκάτῳ ... ἔτει μετ' αὐτήν.

BEFORE 353 B.C.

Demosthenes explicitly says in 24 *Timokrates* 24 that the law about the number of ekklesias, along with other rules, had prevailed for πολὺν ἤδη χρόνον. M. H. Hansen, *The Athenian Assembly* (1987) 22, writes:

[16]R. K. Sinclair, *Democracy and Participation in Athens* (Cambridge 1988) 67, writes, "the denial by Hansen 41–3 of the possibility of additional (emergency) meetings is difficult to accept. Nor is it clear that we should deduce from Dem. 24.21, 25 (indicating three meetings in the first prytany) that in 353/2 only three meetings were held in each prytany (see Hansen and Mitchel (1984) 13–19)."

> I suspect that the Athenians, in the second half of the fifth century and perhaps in the first half of the fourth century as well, held no more than ten obligatory *ekklesiai* in a year, i.e. one *ekklesia kyria* every prytany, and that other meetings, which may have been frequent, were optional.

For a period in the fifth century, Athens was the mistress of a mighty empire, the busiest city in the world, whose embassies went out into all lands, and to whom embassies and deputations were constantly coming. The ekklesia had far more business to transact than in the later centuries. Since the area of Attika is about 700 square miles, it would occasion no surprise if the demos enacted some fixed dates for the convenience of citizens who came from the more remote demes. Dover, in his commentary on Thucydides 6.8 (*HCT* 4.223), postulates that there were three ekklesias at the time of the decision in 415 B.C. to send a fleet to Sicily. At the time of the Five Thousand, Thucydides (8.97.1–2) says the oligarchs held many ekklesias at frequent intervals (Loeb tr.):

> However this may be, after the announcement of their defeat the Athenians, in spite of their difficulties, manned twenty ships, and also called meetings of the assembly—one immediately, which was called to meet on that occasion for the first time in the place called Pnyx, where at other times also they had been wont to meet. At this meeting they deposed the Four Hundred, voted to deliver the management of affairs to the Five Thousand (and all who could furnish themselves with a hoplite's outfit were to belong to this number), and decided that no one was to receive pay for any office, the offender to be accounted accursed. Other meetings also were held later, at frequent intervals, as a result of which they appointed supervisors of the laws and voted the other measures pertaining to their form of government.[17]

[17] In this period, there were ekklesias, not only on the Pnyx, but at Kolonos (8.67.2), Peiraieus (93.1), and one appointed for the theater of Dionysos (93.3).

The phrase μετὰ τοῦτο ἡμέρᾳ πέμπτᾳ ἐκκλησία ἐγίγνετο (Le cinquième jour après, De Romilly) in Thucydides 6.8.3 might suggest that the πρόπεμπτα (πέντε [a cardinal] ἡμερῶν) rule was observed in the fifth century. In the fragmentary Hyperboulos decree (*IG* I² 95; I³ 85), the boule is to make a probouleuma and bring the matter to the ekklesia within five days: προβουλεύσει hε βο]λὲ περὶ τούτον πέν[τε ἐμερôν. Grote long ago (XI.390 of the Second London edition, Chapter 89) observed, "it seems to have been common practice with the Athenians, about peace or other special and important matters, to convene two assemblies on two days immediately succeeding ... προγράψαι τοὺς πρυτάνεις ἐκκλησίας δύο κατὰ τοὺς νόμους." From the evidence of Thucydides 1.44, 3.36.4, and 5.45.4–5.46.1, all illustrating ekklesias on successive days about important matters, it is easy to believe that this law was in force in the fifth century.

A phrase ordering the prytaneis to make an ekklesia (τὴν ἐκκλησίαν ποιεῖν) occurs in *IG* I³ 85.9 and 93.17. In *IG* I³ 61.54, requests made by the cities of the Empire are to be discussed by the ekklesia after sessions (of the boule) held in the dockyards. Since instructions were given, these seem to be special sessions.

Reference is made to κυρία ἐκκλησία in at least two texts. In *IG* I³ 41.37, we have τῆι κυρίαι ἐκκλησίαι without context, as printed in this corpus. In an amendment to a decree ordering that some public works for the water-supply should be executed at the least expense, we have the text as printed in *IG* I³ 49.8–10:

λιγίστον χρεμάτο[ν . . . . 21 . . . . , τὸς πρυτάνες hoì ἂν λάχ]-

οσι πρôτοι πρυτανεύ[εν . . . . 23 . . . . ἐν τêι πρότει τôν]

κυρίον ἐκκλεσιôν πρôτ[ον μετὰ τὰ ιερὰ . . . . 21 . . . . ἀγα]-

For the lacuna in line 9, Hiller following Kirchner in *IG* I² 54 reads [δôναι περὶ αὐτôν τὸν φσêφον ἐς τὲν πρότεν τôν].[18] Wilhelm in *JHS* 68 (1948) 128 would restore [προσαγαγêν τὸν ἀρχιτέκτονα ἐν τêι πρότει τôν]. P. J. Rhodes, *The Athenian Boule* (Oxford 1972) 227, writes that he does not think that lines 8–10 "need imply that there was more than one κυρία ἐκκλησία, or even more than one regular ecclesia, per prytany in the 430s." Since the prytaneis of a phyle were in office for only one prytany, one may argue that the expression is to the first of several kuria ekklesias of that prytany. The phrase ἐν τêι πρότει is restored at the end of line 9 by almost all editors, apparently because of the enigmatic πρôτοι at the beginning of the line.

I find it easy to believe that Demosthenes' πολὺν ἤδη χρόνον extended well back into the fifth century. The reader who consults the two-volume *IG* I³ will realize how much business was transacted by the ekklesia before 403 B.C., and the ratio of preserved documents must be only a small fraction. The prytany calendar was the predominant calendar used in these documents. References to the festival calendar are relatively few; see *ZPE* 128 (1999) p. 90 n. 13 for the names of months in inscriptions earlier than the Choiseul Marble near the end of the fifth century. That the ekklesia held only one ordinary meeting per prytany, I find difficult to consider. One may focus on other factors and speculate differently. B. B. Rogers, for example, devoted three pages of his Introduction (xxvii–xxix) to the *Acharnians* to argue that the dole was a primary factor and the scholiasts' position reflects the period before Athens lost her empire and Aristotle's that of the date of the *Ekklesiazousai* and a later period:

> We know that after the disastrous termination of the War her citizens became quite listless and cared no longer to attend the Assemblies. Nor was the obol which, on the resolution of Agyrrhius, it

[18]Hiller proposed the same phrase in the small fragment, *IG* I² 33.

> was resolved to pay each citizen for each attendance sufficient to overcome their apathy. It was only when the dole was, by the same demagogue, raised to three obols that they again flocked in any numbers to the Assemblies in the Pnyx. See Ecclesiazusae 183–8, 300–10, Plutus 171, 329. And nothing can be more probable than that, to secure a more frequent payment of the dole, the ἐκκλησίαι, instead of being three in every thirty days, should have been increased to four in every thirty-five or thirty-six days; etc.

R. A. De Laix, *Probouleusis at Athens* (1973) 117, writes:

> It has been suggested that it was Cleisthenes who first provided for a set meeting once in each prytany, though no real evidence can be adduced. The term kuria ecclesia first appears epigraphically *ca.* 446/5 (?) in *IG* I^2 42, 22, and later in a literary text in Aristophanes, *Acharnians*, 19. Hignett holds that the use of the adjective kuria is meaningful only when used to differentiate the kuria ecclesia from the other three meetings in a prytany. If so, the system of four assemblies per prytany would date from at least the mid-fifth century. It might be added in support of this view that the *Ath. Pol.* makes no mention of an earlier, different system, as one might expect if the change had been relatively recent.

Whatever theory is evolved from the little testimonia we have must take into consideration that from the middle of the century on the ekklesia handled far more business than in any later period. W. T. Loomis, in *Wages, Welfare Costs and Inflation in Classical Athens* (1998) 22, says that pay for ekklesiasts and prytaneis was introduced to obtain a quorum.

PRYTANY CALENDAR VS. FESTIVAL CALENDAR

At the beginning of this chapter, I grouped together sources which reported that there were four regular ekklesias per prytany, one being kyria (Aristotle, Harpokration, Pollux, Souda, Maximus Planudes), and those which reported κατὰ μῆνα τρεῖς ἐκκλησίας, all being kyria (Photius; Scholia to Aristophanes, Demosthenes, and Aischines; Souda; and Ulpian). The only source which would reconcile the two sets of conflicting testimonia is the following passage in Pollux 8.115 (Bethe):

> πρυτανεία δέ ἐστι ὁ χρόνος ὃν ἑκάστη φυλὴ πρυτανεύει· καὶ ὅτε μὲν δέκα ἦσαν, πλείους ἑκάστῃ φυλῇ αἱ ἡμέραι, ἐπεὶ δὲ δώδεκα ἐγένοντο, ἑκάστη φυλὴ μηνὸς πρυτανείαν ἔχει.

The date would be 307/6 B.C. As with the statement in Aristotle *Ath. Pol.* 43, this rule makes no allowance for the period of the better part of a century when in an intercalary year there were 13 months, but 12 prytanies. In the long 8.95–96 passage s.v. πρυτάνεις for four ekklesias, Pollux makes no reference to a change in the distribution of ekklesias.

The lexicographer Pollux, dated in the time of Commodus (180–192), wrote an *Onomastikon* which in its original form has perished, but extant MSS. are derived from four incomplete copies descending from an early epitome possessed (and interpolated) by Arethas, archbishop of Caesarea ca. A.D. 900. See Bethe's Praefatio.[19] For Book VIII, ὀνόματα δικαστικὰ καὶ πολιτικά, see Bethe, *RE* 10.1 col. 778. According to J. E. Sandys, *A History of Classical Scholarship* I.327, his primary sources were the lexicons of Didymus, Tryphon, and Pamphilos. The Pollux passage was treated in Chapter 4 of *The Calendars of Athens* (1947. pp. 78 ff.) in the context of an effort to show the difference between prytany and festival dates in an ordinary year of twelve phylai. Plutarch (*Dem.* 8.3) says that Demetrios Poliorketes appeared off Peiraieus on πέμπτῃ φθίνοντος Θαργηλιῶνος (XI.26). He then ran a trench and palisade round Mounichia (9.2), and sailed against Megara which he captured (9.5). He returned to Mounichia and demolished the fortress (10.1). At last he made his entry into the city and gave the Athenians their

[19]See also Tolkiehn, *RE* s.v. Lexikographie (1925) 2457–2458.

ancient form of government. Ferguson (*Hellenistic Athens* 95–96) estimated that his final entry was about two months after his appearance off Peiraieus. Thereafter, it must have taken considerable time to work out the reassignment of about 25 demes and other administrative details, as well as the repair of the walls. *IG* II^2 466, although lacking a prescript, has been assigned to 307/6 because of the historical context. The decree validates a covenant with Tenos. It refers to Mounichia in line 5 and to Demetrios and Antiochis in line 7. Although these lines defy restoration, the three names suggest that the inscription must be after the fall of Mounichia and Demetrios' entry. Cf. Beloch, IV^2.1.151; IV^2.2.244. As Ferguson notes (p. 96 n. 2) several sources (Parian Chronicle, Diodoros, Dionysios Hal.) represent the arrival of Demetrios in Athens *during* the archonship of Anaxikrates, i.e. after the beginning of 307/6 B.C. The interesting feature is that the inscription in line 45 refers to the Boule of Five Hundred. A. G. Woodhead, who, in *The Athenian Agora Vol. 16 Inscriptions: The Decrees* (1997), slavishly follows the reconstructions of the two calendars by B. D. Meritt, writes (p. 169), "The last-named contains a reference to the 'Council of the Five Hundred' and could be assigned to the last months of 308/7, after the liberation of Athens, rather than to 307/6." This can hardly be reconciled with the record in Plutarch. Moreover, Diodoros has the capture of Mounichia and Demetrios of Phaleron's flight from the city within the year 307/6: 20.45.3–4. Dionysios of Hal. *Deinarchos* 3 even quotes a fragment of Philochoros (328 frg. 66) which has the capture of Megara, the taking of Mounichia, and the restoration of democracy within the archonship of Anaxikrates. Since Philochoros was born c. 340 and held the official title of μάντις καὶ ἱεροσκόπος in 306, it seems unlikely that he was wrong. However, the literary evidence is discredited by Jacoby in his commentary on the Philochoros fragment on the basis of the restoration of the calendar in *The Chronology of Hellenistic*

Athens (1940) Chap. 1, in which *IG* II² 456 was called "the cornerstone of a sound chronology." The inscription in question is dated in Prytany V.20+ and the month Maimakterion (V) ?. The day of the month and the first ordinal of the prytany date are restored in a stoichedon 41 text. Meritt returned to the calendar in several texts with different reconstructions of the calendar, always affirming, "Their number (prytanies) was already fixed as 12 before the year began": *Athenian year* 178 n. 19, a statement in which I have concurred. Woodhead (168), of course, adopts Meritt's latest version, which has 30 days assigned to the first prytanies, a calendar attested nowhere else. Of the nine prescripts from this year (plus *IG* II² 1487.53–54 and Plutarch *Mor.* 852A), only two (*IG* II² 458 and *Chronology* p. 8) are sufficiently preserved to give positive equations. The alleged decree in Plutarch *Mor.* 852A gives Antiochis the sixth prytany, if Schömann's emendation of ἕκτης is correct.

Meritt wrote in *Hesperia* 5 (1936) 204 n. 1, "In spite of the numerous documents preserved for the year 307/6 there is only one equation before the tenth prytany where the calendar implications do not depend on restoration. This is found in I.G., II², 456 and shows that some day in the twenties of the fifth prytany fell in Maimakterion," where the day must also be restored. Kirchner, Broneer, and West have proposed calendric schemes for the year, Dinsmoor two, and Meritt three, all eight differing and all by experienced students of the calendar. Dinsmoor (*Hesperia* 4 [1935] 306) refers to "the senseless havoc in the equation of *IG* II² 456" and posits (*Archons of Athens* 383–384) "that Maimakterion was written for Posideon," or τετάρτης for πέμπτης, with a choice of two different equations.

I review some of the calendric factors. Leonardos (*AD* 1916 Para. 67–69), from the stone, and I (*AJP* 58 [1939] 220), from a squeeze, independently observed that line 16 of frg. b of *IG* II² 456 should read that a boule of 500 was in office, as was the case in *IG* II^2 466.45. The boule of 600 members had not been created as late as the fifth prytany and the fifth month. yet the new Antigonis is on the stone (*IG* II^2 458) as being in prytany in the seventh (reading [ἐβ]δ[όμης] instead of Kirchner's [ὀγ]δ[οίης]) prytany in the month of Gamelion 1 (VII). We know from *IG* II^2 1487.53–54 that there was a second Gamelion, a text often overlooked.[20] A boule of 600 was apparently created during the course of the sixth prytany held by Antiochis. Finally, we know from *IG* II^2 458 that the archon tampered with the calendar in Gamelion 1 (VII).

Rather than rejecting the literary evidence, including Philochoros, I suggest, following Pritchett/Meritt, that the year began as ordinary with a boule of 500. Demetrios Poliorketes had landed on Attic soil in Thargelion (XI) 26 of 308/7, as Plutarch says, taken Peiraieus, and walled off Mounichia. With all of Attika freed except Mounichia, this was a signal for the restoration of the democracy by the beginning of 307/6. The determination to create the two Makedonian phylai was made at that time, but the process of the redistribution of 25 or so demes, with some of the demes such as Phyle and Lamptrai being divided, required the better part of 6 months. Such assumptions have been made earlier by students of the calendar, except I would propose that in the 'cornerstone' inscription, *IG* II^2 456 (stoichedon 41), instead of Kirchner's restoration in line 4

[ριῶνος ἕνηι καὶ νέαι, δευτέραι] καὶ εἰκοστῆι τῆς [πρυ-]

and Meritt's [ριῶνος ἕνηι καὶ νέαι, τετράδι] καὶ εἰκοστῆι τῆς [πρυ]-

the reading of [ριῶνος δεκάτηι ὑστέραι, μιᾶι] καὶ εἰκοστῆι τῆς [πρυ]-

Maimakterion (V) 21 = Prytany V 21. The two calendars were running pari passu for an ordinary year of 12 phylai, as Pollux

[20] A good collection of the numerous references to the archon Anaxikrates is given by Dinsmoor (*The Athenian Archon-List in the Light of Recent Discoveries* [New York 1939] 27), but he omits this important reference.

has it. Posideon (VI) 1 would equal Prytany VI 1 and c. 149th day of the year. Working backwards from the one completely preserved equation of the year in *IG* II^2 458, we have Prytany VII.21 = Gamelion 23^3, or Prytany VII.18 = Gamelion 22, and, if there was no tampering with the festival calendar, Prytany VII.1 = Gamelion 4. Prytany VI had been lengthened by three days. Presumably all subsequent prytanies were also lengthened for an intercalary year of 384± days, and a boule of 600.

At this stage, the archon played havoc with the calendar in a way that no modern can adequately explain; thus we have at least eight different reconstructions. Meritt, in his latest reconstructions (*Hesperia* 32 [1963] 43–437; 33 [1964] 13–15), assumed 12 embolimoi days were added in Elaphebolion (IX), 11 before the 9th and 1 after the 30th, which were "compensated" (i.e. subtracted) during Mounichion (X). Thus, his festival month Elaphebolion (IX) has 42 days and his Mounichion (X) 17 days. Dinsmoor alleged scribal errors in dates and could cite parallels. We have no idea why the archon began the intercalation of days on Gamelion 22, and the fact that *IG* II^2 458 was passed on a second intercalated Gamelion 22 does not mean that the intercalation stopped with Gamelion 22^3. We do not know how many days were in Gamelion I nor why the year was changed from ordinary to intercalary by the addition of Gamelion II. It is meaningless to claim probative value for restorations of texts such as *IG* II^2 459, which has a few letters of three lines preserved in a text in which the length of line cannot be proved.[21] Dow (*HSCP* 67 [1963] 72) remarked, in reviewing Meritt's restorations in this period, "If so many variables can be so easily brought into play, the whole business is a deuces-wild game, and the game is a game played with bluffing."

[21]Compare Mikalson's first paragraph on his p. 126 for restorations for a single day in 307/6. Oddly, Broneer (*Hesperia* 2 [1933] 400) claimed the *IG* II^2 459 text was wrong.

My purpose has been only to show that the pattern of prytanies in the first five months might conform to one well-attested elsewhere. That one scholar without new evidence can produce three different charts for the calendar of the year illustrates the futility of generalizations about the latter part of the year 307/6.

In the period of the two timocracies after the Lamian War, we cannot be sure about the number of ekklesias per prytany, but I imagine, without any evidence, that the archon of 307/6 at the beginning of the year was continuing a schedule already in existence. Demetrios of Phaleron, a pupil of Theophrastos, had limited "the franchise to possessors of one thousand drachmae": Ferguson, *Hellenistic Athens* 39. His timocracy followed an even more restrictive one in which Phokion was accused of being responsible for the κατάλυσις τῶν νόμων (Diodoros 18.66.5). The franchise was limited to possessors of two thousand drachmai (Diodoros 18.18.5). Plutarch (*Phokion* 28.4) says twelve thousand were deprived of their franchise, but adds that Phokion managed the city νομίμως (29.4). Pay for attendance had been withdrawn. S. V. Tracy, *Athenian Democracy in Transition* (Berkeley 1995) devotes a chapter to the inscriptions of Demetrios of Phaleron, writing (p. 36), "One of the most notable facts about the period 317–307, the decade of Demetrios' rule, is that almost no decrees of the assembly can be assigned to these years. Indeed there are now only two certain ones, *IG* II^2 450 and 453." He adds that J. Morgan will argue that the fragment *IG* II^2 585, which lacks the name of the archon but has a calendar equation which by Kirchner's restorations suggests an intercalary year, belongs to the year 314/3. Meritt and Dinsmoor (*Hesperia* 32 [1963] 4) restored it for an ordinary year.

I noted above that three of the sources (scholia to Aristophanes and Aischines and the Souda) which reported three ekklesias per month stated that all were called kyria. Two kyria ekklesias were held on different days in the same prytany (XII) and month (XII) in 303/2: *IG* II² 494 and II² 498 with add. Any suggestion that this is to be explained because there was pressing business at the end of the year would ignore the fact that another ekklesia occurred later in the prytany. In one example where we have two decrees passed on the same day, one gives the designation ἐκκλησία, the other ἐκκλησία σύγκλητος: *IG* II² 911 and *Hesperia* 5 (1936) 429. Hence, since the majority of decrees give only ἐκκλησία, one may argue that some of these were kyria, but the word was omitted. We have decrees in which the name of the month is omitted or the date in the month (e.g. *IG* II² 360.3; 464.5). Scribes and masons tolerated omissions, which presumably were not found on official copies in the bouleuterion. Nonetheless, I find no recent scholar in the literature who supports the theory of three kyrias or even mentions it. If the rule in Aristotle applied, attendance at a kyria ekklesia brought higher remuneration than at other sessions, although we know that the timocracy of Demetrios abolished payments.

The question may be raised whether κατὰ μῆνα was used loosely for κατὰ πρυτανείαν. This, I believe, was the theory of M. H. Hansen in his early studies. Our literary sources speak only of ordinary years, when in the period of twelve months there was an equal number of prytanies. Aristotle refers only to an ordinary year for his prytany divisions. A prime example is in Plato *Laws* VI, where he tells us in 767C that in his ideal state the new year begins with the month next after the summer solstice, but in 758D that his boule, which controls his ekklesias, shall be in office for the twelfth part of the year. Twelve prytanies were an ideal method of equating the prytany with the festival calendar in an ordinary year, but on average seven out of nineteen years were intercalary; so the system breaks down.

In the κατὰ θεόν (and κατ' ἄρχοντα) calendar equations in the period of twelve phylai which give the prytany dates, the kata theon and the prytany dates are usually the same in ordinary years, but there are two examples where the kata theon dates in the second century show a difference of one day from the prytany dates: *Chronology of Hellenistic Athens* p. 121 (IX 27 k.t. = IX. 28) and *Hesperia* 26 (1957) 72–73 (IX.21 = IX 22). The prytanies were in office according to their own calendar. As we shall discuss below, this was to be expected since the length of prytanies was determined in advance, whereas that of the lunar month was not established until the end of the month. Needless to say, equality would not prevail in intercalary years of thirteen months. Pollux' statement ἑκάστη φυλὴ μηνὸς πρυτανείαν ἔχει is not literally true.

There remains the question of any schedule of fixed dates for the prytany calendar. Two of the sources cited above attesting three ekklesias per prytany say they met on the 1st, 10th, and 30th (scholion to Aristophanes *Ach.* and the Souda), one says they met on the 11th, around the 20th, around the 30th (scholion to Demosthenes 24), and one on the 11th, 20th, and 30th (Ulpian). The Souda is of great importance, but has many corruptions. No one familiar with Attic decrees would endorse its calendar of ekklesias, or that of the *Acharnian* scholion. E. M. Harris, in his 1991 *AJP* article (328 ff.), written in support of the Demosthenic scholion, gives catalogues of days of meetings of the ekklesia in inscriptions in terms of the festival calendar (only). In the entire article, no reference is made to intercalary years, which on average would occur in 7 out of 19 years. This means that for Harris in an intercalary year of twelve phylai, three prytanies had only two regular ekklesias. There are gaps in Harris' tables. In his tables on pp. 328 and 337, he writes for Day 6: "There is no indisputable evidence for a meeting on this day." No. 37 of O. Kern, *Die Inschriften von Magnesia am Maeander* (1900), which has lost the archon's name, was set up at the cost of the tamias and the plural board of administration at a time when Chalkis was in possession of Philip V, so before 199 B.C. (see Kern's notes to 37 and 47), and belongs to the period of 13 phylai.[22] It is dated in the fifth prytany: Πυανοψιῶνος ἕκτει ἱσταμένου, [ἑβ]δόμε[ι] τῆς πρυτανείας.

PYANOPSION (IV) 6 = PRYTANY V 7

In *Calendars of Athens* 90, I regarded the equation as one for an intercalary year. Dinsmoor (*Athenian Archon-List* 234) assumes πέμπτηι was written for τετάρτηι. Again, in *IG* II^2 1014, we have what from the spacing in line 7 is surely a decree of the boule and demos: Pyanopsion (IV) 6 = Prytany (?) 5. Harris, as he indicates on p. 328, omits all decrees of the boule (ἔδοξεν τεῖ βουλῆι). The prytaneis of a phyle were often praised by decrees of both the boule and the demos: Meritt and Traill, *The Athenian Agora* XV.5–6. A boule of 500, 600, or 650 was using the same calendar as in the demos decrees. Decrees of the boule, of course, provide no evidence for the dates of the ekklesias, but they show that these large bodies met on days on which Harris finds no evidence for a meeting of the ekklesia: Maimakterion (V) 6 (*Agora* XV no. 194); Skirophorion (XII) 3 (*IG* II^2 847); Metageitnion (II) 2; Metageitnion (II) 4 (Dow, *Prytaneis* no. 84); etc. There are a number of these.[23]

With regard to the two sources which say the Athenian ekklesia met on the 20th or around the 20th, it is noteworthy that Harris' tables show only one example of an ekklesia on that date. It was not customary to hold a meeting on festival days, and these did not fall on the same day in the several prytanies and months. Hence it was customary to give a five days' notice of the ordinary

[22]See Habicht, *Athens from Alexander to Antony* (1997) 193.

[23]According to Harris' two tables, there were more decrees dated on Skirophorion (XII) 30 (11) than on any other single day. This proves only that, like many administrative bodies, ekklesias made decisions at the last moment. Indeed, there are more decrees in his tables dated in the month Skirophorion (25) than in any other one, although the month Elaphebolion (IX), with its Dionysiac festival, runs a close second.

meetings in a πρόγραμμα which contained also the agenda of the meeting. See Aristotle *Ath. Pol.* 43.4: προγράφουσιν δὲ καὶ τὰς ἐκκλησίας οὗτοι (= οἱ πρυτάνεις); Photius πρόπεμπτα; *Lexica Segueriana* (1.296.8, Bekker):

> Πρόπεμπτα: τὸ πρὸ πέντε ἡμερῶν τῆς ἐκκλησίας προγράφειν, ὅτι ἔσται ἐκκλησία. εἰ τύχοι, εἰ ἔδει ἐκκλησίαν γενέσθαι τῇ δεκάτῃ, προέγραφον οἱ πρυτάνεις ἀπὸ τῆς πέμπτης ὅτι ἔσται.

Since we believe that the first ekklesia of the year fell on Hekatombaion 11 in the fourth century, as inferred from Demosthenes 24.20, 26, it is probable that other prytanies would normally devote their first days to a πρόγραμμα. In several inscriptions, we have the formula ὅταν αἱ ἡμέραι αἱ ἐκ τοῦ νόμου ἐξήκωσιν, χρηματίσαι περὶ - - -: *IG* II^2 652, 667, 673, etc.

The whole concept that the Athenian ekklesia met according to fixed (or approximate) dates in the festival calendar is invalid. Certainly, no epigraphist is going to offer revised restorations in conformity with such a theory, as happened for some dates after Mikalson reminded us that important festival dates were avoided.

SYNKLETOS EKKLESIA

In connection with the problems of the number of ekklesias and whether they were tied to the prytany or the festival calendar is that of the synkletos ekklesia. It is noteworthy that the references in the scholia to this ekklesia occur in passages quoted above (Aristophanes *Ach.* 19; Aischines 1.60; Demosthenes 18.73, 19.123, 24.20), all of which state that there were three ekklesias per (festival) month. In addition, as noted, Harpokration, in contrast with his statement that there were four ekklesias per prytany, says s.v. σύγκλητος ἐκκλησία that ekklesias were "habitual and by month" with the exception of the synkletos (Keaney):

(Σ 59) Σύγκλητος ἐκκλησία· τῶν ἐκκλησιῶν αἱ μὲν ἐξ ἔθους καὶ κατὰ μῆνα ἐγίνοντο· εἰ δέ τι ἐξαίφνης κατεπείξειεν ὥστε γενέσθαι ἐκκλησίαν, αὕτη ἐκαλεῖτο σύγκλητος ἐκκλησία· Δημοσθένης ἐν τῷ Κατ' Αἰσχίνου (19.122).

The same text is found in the *Souda* (1292), which as Adler notes was taken from Harpokration. Likewise in the *Ety. Mag.* In his 1853 edition, Dindorf suggests that Harpokration found this statement as a gloss on the spurious decree in a text of Demosthenes (18.37).

In our collection of sources which specify four meetings per prytany (above, pp. 186–187), Maximus Planudes also wrote that the synkletos ekklesia was an additional meeting for a matter of great concern. Thus all the lexicographical sources which mention the synkletoi, including Harpokration, the Souda, *Ety. Mag.*, Pollux, Planudes, and several scholiasts are in agreement that this extraordinary meeting was in addition to the regular ekklesias. In a position which belongs to the eccentricities of scholarship. H. M. Hansen, in maintaining his view, repeatedly advocated, that the synkletos ekklesia was never an extra meeting, writes (*The Athenian Assembly* [1987] p. 29), "The only sources to be adduced <for an extra meeting> are some muddled notes in scholia and lexica." Contradictorily, although Hansen is a firm supporter of the brief and incomplete account of ekklesias in *Ath. Pol.* 43.2–6, much of his presentation of ekklesias is based on these same sources.

There are two spurious decrees inserted into Demosthenes 18 *Crown* which read συγκλήτου ἐκκλησίας ὑπὸ στρατηγῶν καὶ πρυτάνεων καὶ βουλῆς γνώμῃ (*Crown* 37) and ἐκκλησίας συγκλήτου ὑπὸ στρατηγῶν (*Crown* 73). As noted above (p. 129), there is a consensus of opinion that these were drafted in the second century. The context of the apocryphal decree in Demosthenes 18 *Crown* 37–38, studied by Grote (XI.422–423), Schaefer (II^2 279), and Goodwin (*Demosthenes De Corona* 263–264), is the ekklesia at Peiraieus in which the alarming news of the surrender of the Phokians at Thermopylai was brought by Derkylos, who came directly into the meeting in progress. The people were holding an assembly on matters connected with the docks and arsenal. The date was Skirophorion (XII) 27 347/6 = Dem. 19.60 (τετράδι φθίνοντος). The shock to the public was prodigious. In their panic and on the motion of Kallisthenes, the people voted to remove the women and children from the country into protected towns or fortresses, to put the Peiraieus and the forts in a state of defense, and to hold the coming festival of Herakles, held in the country, within the city walls. See in particular Demosthenes 19.125.[24] The context of the second decree is acknowledgment that Philip had broken the peace in the late summer of 340; see Goodwin 283. R. G. Ussher, in his commentary on Aristophanes *Ekklesiazousai* (1973), lines 551–553, cites the σύγκλητος ἐκκλη-

[24]As to the Herakleia, Mathieu in the Budé comments, "Enfin le décret omet la mesure que Démosthène indique ailleurs (*Amb.* 86 et 125) comme la plus caractéristique: la célébration *intra muros* des Héracleia (de Marathon)." Earlier, Grote (XI.423 n. 2) commented on the decree as so spurious as to lack all authority, referring to Droysen and other critics. The interesting feature is that the festival was not canceled; only the site was changed. There were only two days left in the month (and year), if hollow, or three if Skirophorion was full. According to L. Deubner, *Attische Feste* (1932) 227, this festival was a penteteric one, and hence probably very elaborate. The approximate site of the Marathon sanctuary has been fixed by E. Vanderpool's discovery of inscriptions. Harpokration writes (text of Keaney), "(H 14) Ἡεράκλεια· Δημοσθένης ἐν τῷ Κατ' Αἰσχίνου (19.86, 125). πολλῶν ὄντων τῶν κατὰ τὴν Ἀττικὴν Ἡρακλείων, νῦν ἂν ὁ Δημοσθένης μνημονεύοι ἤτοι τῶν ἐν Μαραθῶνι ἢ τῶν ἐν Κυνοσάργει· ταῦτα γὰρ μάλιστα διὰ τιμῆς εἶχον Ἀθηναῖοι." The Herakleia in Kynosarges in the deme Diomeia was held in Metageitnion (II): A. Mommsen, *Feste der Stadt Athen* (1898) 160–161; H. W. Parke, *Festivals of the Athenians* (1977) 51. R. Parker in *Tria Lustra: Essays Presented to John Pinsant* (1993) 25–26, would place the site north of the city on the basis of *IG* II^2 1247. For the standard view, see J. S. Traill, *Hesperia Suppl.* XV (1975) 39; J. Travlos, *Pictorial Dictionary* (1971) 340–341.

σία in Demosthenes' text (18.73) as a parallel for Praxagora's specially summoned meeting. See also B. B. Rogers' commentary (1919) on the lines.

Pollux (8.116), in the second century, refers to synkletos ekklesia as being called on a sudden for a matter of great importance. He adds that it is also called κατακλησία because they summon people living in rural areas: σύγκλητος ἐκκλησία, ἣν ἐξαίφνης ἐποίουν μείζονος χρείας ἐπιλαβούσης. ἐκαλεῖτο δὲ καὶ κατακλησία, ὅτι καὶ τοὺς ἐκ τῶν ἀγρῶν κατεκάλουν. Three centuries later Hesychius s.v. κατακλησία abbreviated the account: κατακλησία· ὅταν τῶν μειζόνων τι δέῃ πρᾶγμα ἐπικρῖναι, κατακλησίας ἐποίουν τῶν ἐν τοῖς ἀγροῖς οἰκούντων. The grammarian Ammonius, in i/ii A.D., explains (p. 47, L. C. Valckenaer[2] Leipzig 1822):

> Ἐκκλησία καὶ κατάκλησις διαφέρει. ἐκκλησίαν μὲν γὰρ ἔλεγον οἱ Ἀθηναῖοι τὴν σύνοδον τῶν κατὰ τὴν πόλιν· κατάκλησιν δὲ, ὁπότε καὶ τοὺς ἐκ τῶν ἀγρῶν συνεκάλουν πρὸς ἐπίσκεψιν μείζονα τῶν πραγμάτων·

The term κατάκλησις never turns up in epigraphical sources. Aristotle *Ath. Pol.* 43.4 wrote προγράφουσιν δὲ καὶ τὰς ἐκκλησίας οὗτοι (= οἱ πρυτάνεις). Photius explains that the πρόπεμπτα, five days (πέντε ἡμερῶν) notice of ordinary meetings, was ἵνα καὶ οἱ ἐν τοῖς ἀγροῖς συνέλθωσι. Hansen (e.g. *The Athenian Assembly* 29), apparently under the impression that all cardinals and ordinals in Greek are to be reduced by one in modern count, mistranslates the adverb as "four days' notice", but see LSJ, Photius, Demosthenes 43 [75], *IG* II^2 1237.61.

One notes that Pollux and Harpokration apply the word ἐξαίφνης 'on a sudden' to a meeting called synkletos. This is also true of Demosthenes. The context of the Demosthenes passage with ἐξαίφνης in 19 *False Legation* is that the orator argues that treachery is proved by Aischines' conduct on the appointment of the third embassy (19.121–133), which ultimately set out about the 21st of Skirophorion (XII) 347/6 (W. Goodwin, *Demosthenes De Corona* 263; A. Schaefer, *Demosthenes* II[2] 276–278; Grote XII.422). Meanwhile, letters had arrived from Philip inviting the Athenians to join him, but, as Demosthenes shows, really sent to prevent the Athenians from marching out to aid the Phokians holding Thermopylai. There was danger in the invitation being accepted, and all sorts of discussion were going on in the agora. In this context Demosthenes says they were afraid that a synkletos ekklesia might suddenly be convened and they would hear the truth from him and the Phokians might be saved.

Relevant texts in Demosthenes and Aischines are the following.

Demosthenes (19 *False Legation* 123):

> ἐφοβοῦντο δὴ μὴ σύγκλητος ἐκκλησία γένοιτ' ἐξαίφνης, εἶτ' ἀκούσαντες ὑμεῖς ἐμοῦ τἀληθῆ ψηφίσαισθέ τι τῶν δεόντων ὑπὲρ τῶν Φωκέων, καὶ τὰ πράγματ' ἐκφύγοι τὸν Φίλιππον.
>
> (Loeb tr.: They were afraid that an extraordinary meeting of the Assembly might suddenly be convened, and that then, on hearing the truth from me, you might adopt some acceptable resolution in favour of the Phocians, and that so Philip might lose control.)

Aischines (2 *Embassy* 72) likewise uses the term once in a passage which proves conclusively a distinction between extraordinary and regular ekklesias:

> πλείους δὲ ἐκκλησίας συγκλήτους ἠναγκάζεσθε ἐκκλησιάζειν μετὰ φόβου καὶ θορύβου, ἢ τὰς τεταγμένας ἐκ τῶν νόμων·
> (Loeb tr.: And the special meetings of the assembly which you were forced to hold, in fear and tumult, were more in number than the regular meetings.)

Finally, the word occurs in Aristotle *Pol.* 3.1275b, where W. L. Newman in his commentary states:

> **συγκλήτους**, such as, for instance, the 5000 at Athens, whom the 400 were to call together whenever they pleased (Thuc.

8.67.3). The contrast drawn implies that it was of the essence of an ἐκκλησία to meet at regular intervals, and not merely when the authorities of the State chose to convoke it. Compare ...

For purposes of comparison, we may turn to the specially convened synkletoi of the Achaian League. The word σύγκλητος is used many times by Polybios, especially for Rome, but also for the Achaian League. The nature of this latter assembly has received treatment from Aymard, Larsen, Cary, and others, as reported in Walbank, *HCP* 1 p. 244; 3 (1979) 399, 413–414. In the last passage, Walbank writes:

> To sum up: the Achaeans had a Council and a primary assembly which met at four annual *synodoi* right down to 146. From the time the Romans appeared on the scene, towards the end of the third century, certain delicate issues were removed from the competence of the regular *synodoi* and reserved for discussion and decision at a specially convened *syncletos* with a single item on the *agenda*; such *syncletoi* would normally consist of the Council and assembly, but (as in 168) exceptionally, following a decision either of the *synodos* or of the magistrates (we are not told which), could have a different composition. Membership of the assembly was open to all men of military age; membership of the Council was probably restricted to men of thirty and over.

Mogens H. Hansen and F. Mitchel summarize their position in *SO* 59 (1984) 17:

> (a) that four meetings in a prytany was not only a minimum but also a maximum, and (b) that an ἐκκλησία σύγκλητος was not an extra meeting, but one of the four obligatory meetings summoned at short notice in an emergency, or by orders of the *boule* (and not by the *prytaneis* on their own initiative).[25]

[25]As recently as 1993, Mogens Herman Hansen (*AJP* 114.99) wrote in the opening sentence of his article: "The scholion on Demosthenes 24.20 is the only source in which it is explicitly stated that *ekklēsia synklētos* was

The statements in the lexicographers about four meetings per prytany all derive from Aristotle *Ath. Pol.* 43.3, who is characterizing normal conditions, but by no means giving a constitution governing all procedures. For example, in the same section (43.2), he refers to a calendar for the boule which applies only to ordinary years. He says nothing about intercalary years; yet all would agree that the Athenian calendar was not a sequence of years of 354±1 days. That the rule about ekklesias is not complete is illustrated by the fact that Aischines says (2.68) that κατὰ τὸν νόμον there were two ekklesias to deliberate on the question of peace and (2.58) that ambassadors of foreign states could address the ekklesia. A priori, one would assume that in cases of emergency, an extraordinary ekklesia would be assembled on short notice, and Demosthenes 19.123 (ἐξαίφνης) and Aischines 2.72 (πλείους ἢ τὰς τεταγμένας ἐκ τῶν νόμων) confirm this assumption. At the time of the Five Thousand, the codices of Thucydides read that there were assemblies in quick succession (8.97.2): ἐγίγνοντο δὲ καὶ ἄλλαι ὕστερον πυκναὶ ἐκκλησίαι. For an ekklesia which seems to have been special for the occasion, see Xenophon, *Hell.* 1.7.4.

The crux of Mogens H. Hansen's position is that the synkletos ekklesia must be one of the Aristotelian four. In *GRBS* 20 (1979)

---

an extra meeting of the assembly and not just one of the ordinary sessions summoned in a special way: that is, with shorter notice than usual and/or by a decree of the council or the people, not simply by the *prytaneis* on their own initiative." Not only is the first part of his statement completely false, but there is no ancient authority who defines a synkletos ekklesia as he does in the second part. In his next paragraph, he concedes that this same scholiast "is wrong or, at best, anachronistic" about the rest of the scholion. Hansen's method is eclectic with no consideration of the value of his sources. Finally, his scholiast does not use the word synkletos. It is utter confusion to say that in one text an "ekklesia not ordained (ὡρισμένη or τεταγμένη) by law" is a synkletos, whereas in others it is not. I referred to this sort of scholarship in my *Essays in Greek History* (1994) 27–110.

149–156, he collects examples in epigraphical texts where σύγκλητος (or σύνκλητος) is applied to βουλή and ἐκκλησία. The list has been expanded by S. V. Tracy in *ZPE* 75 (1988) 186–188, who catalogues eight examples from Athens and one from Athenian Delos (144/3 B.C.). The Athenian texts are dated between 226/5 and 168/7. All are honorary inscriptions, which is true of the majority of inscriptions from this late period. There are two inscriptions passed on the same day, the last of the year 169/8, an ordinary one: Skirophorion (XII) ultimo = Prytany XII.29. Kalliphanes of Phyle, who took part in the battle of Pydna on the side of the victors, when Aemilius Paullus defeated King Perseus, was awarded honors for bringing news of the battle: *Hesperia* 5 (1936) 429–430 = L. Moretti, *Iscrizioni storiche ellenistiche* 1 (1967) no. 35.[26] The Athenians probably celebrated a εὐαγγέλια. There is a lacuna in lines 8–10 of the prescript where the names of the president of the proedroi and the symproedroi would have been inscribed. The ekklesia is designated as ἐκκλησία ἐμ Πειραιεῖ. That the ekklesia was a synkletos one we learn from *IG* II² 911, passed on the same day: ἐκκλησία σύ[γκλητος ἐμ Πειραιεῖ στρατη]γῶν παραγγ[ειλάντων].[27] The body of the text is not preserved. The scribe or mason of the Pydna text omitted the word synkletos. That is not exceptional. Many inscriptions tell us only that there was an ekklesia without further qualifications, but the vagaries of secretaries and masons are hard to explain. Other decrees than the main item on the agenda might be passed at the same meeting.

One of the inscriptions with synkletos boule, *IG* II² 897, was for a high official in the service of Ptolemy V: βουλὴ ... σύγκλητος στρατηγῶν παραγγειλάντων. See C. Habicht, *Athens*

[26]Habicht, *Hesperia* 59 (1990) 570 n. 45, writes, "Hansen failed to realize that it was the news about the battle of Pydna that made the generals eager to have a meeting of the assembly called on short notice."

[27]For the text, see S. Tracy, *ZPE* 75 (1988) 187.

from Alexander to Antony (Harvard 1997) 221. Another (*IG* II^2 954), with a similar formula, was a grant of citizenship to an official Pausimachos in the service of King Eumenes: Habicht, *Hesperia* 56 (1990) 567. At a synkletos ekklesia in the theater an honorary decree was passed for Kastor of Alexandria, a high-ranking Ptolemaic official: *IG* II^2 838; Habicht, *Athens from Alexander to Antony* (1997) 178. The synkletos ekklesia of *IG* II^2 945 honored a certain Diodoros, who is described as a friend of King Eumenes II and his brothers; cf. Habicht, *Hesperia* 59 (1990) 570. This decree was passed on the last day of the year. Thus, three of the synkletos meetings were held on Skirophorion (XII) ἕνει καὶ νέαι, which might suggests that they were special sessions.

For the synkletos ekklesia of *Hesperia* 36 (1967) 64, we have no context. In the Delos inscription (*IdeD* 1507 lines 37 ff.), passed in Prytany VIII, the generals convoked a synkletos ekklesia in the theater to ratify honors for the epimeletai of the emporion on Delos. In a small lacuna in the formula, S. Tracy (*ZPE* 75 [1988] 187–188) would restore κυρία for this assembly. The one synkletos inscription which might seem difficult to reconcile with an extraordinary meeting, in this case of the boule, is *Athenian Agora* 15 no. 167 p. 145:

> βουλὴ ἐμ βουλευτηρίωι σύνκλητος στρατηγῶν παραγγειλάντων
> καὶ ἀπὸ βουλῆς ἐκκλησία κυρία ἐν τῶι θέατρωι

The decree, passed in an intercalary month, honors the phyle Leontis and is preserved as far as the ἀναγράψαι formula. The body of the text contains nothing but the usual formula for conducting sacrifices and attending to the συλλογή of the boule and demos, certainly nothing of a military nature ostensibly to warrant the intervention of the generals. However, the ἀρχαιρεσία included the election of military officials (*Ath. Pol.* 44.4), and from the experience of legislative bodies in Washington and my state capital, it is quite possible that what seemed innocent decrees had ulterior motives.[28] The text includes what W. Larfeld (*Handbuch der griechischen Epigraphik* 2.2. [Leipzig 1902] 685 ff.) called the 'Referatformel', stating that it became common at the end of the third century: ὑπὲρ ὧν ἀπαγγέλλουσιν οἱ πρυτάνεις τῆς Λεωντίδος ὑπὲρ τῶν θυσιῶν ... . There are scores of such prytany inscriptions. One may suggest that, as with *Hesperia* 5 (1936) 429–430 and *IG* II² 911, both passed on the same day, the body in question was concerned with other matters than the principal one for which they were assembled. In the texts which have the βουλὴ σύνκλητος phrase,[29] the decrees were adopted as law at

[28]Meritt (*Klio* 52 [1970] 282), for example, speculates that at the synkletos boule called at the behest of the generals (*IG* II² 954a), the decree to make Pausimachos a citizen was for the ultimate purpose to elect him to the generalship at that same ekklesia.

[29]Aristotle (*Ath. Pol.* 43.3) says the boule convened every day except ἀφέσιμος (ἡμέρα). If the rule still applied a century and more later, it would seem that the convening of the boule by the generals signified a matter of some urgency when they appeared and presented the agenda. The power of generals in calling meetings is debated. However, Plutarch *Phokion* 15.1 is explicit (Loeb tr.): "Phocion, fearing that the Boeotians might get early knowledge of the appeal and anticipate Athens in sending help, called an assembly early in the morning and announced to the Athenians the message received from Megara. Then, as soon as the requisite decree had been passed, he ordered the trumpeter to give the signal and led them, under arms, directly from the assembly." The ekklesia passed a decree. Moreover, in five of the examples of σύγκλητος βουλή or ἐκκλησία collected by S. Tracy, *ZPE* 75 (1988) 186–188, we have the phrase στρατηγῶν παραγγειλάντων: *IG* II² 897, 911, 954; *Athenian Agora* 15.167; *Hesperia* 15 (1967) 64. παραγγελία usually means 'command' or 'order'. These five are dated at the beginning of the second century. In Thucydides 2.59.3, we read (Loeb): "And when he (Perikles) saw that they were exasperated by the present situation and were acting exactly as he had himself expected, he called a meeting of the assembly—for he was still general—wishing to reassure them, and by ridding their minds of resentment to bring them to a milder and less timorous mood. So he came forward and spoke as follows:". Here the meeting is called σύλλογος, but in 2.60.1 is referred back to as

statutory meetings of the ekklesia, termed κυρία or ἀρχαιρεσία (for the election of officials).[30] Thus, in the case of *IG* II^2 954, the ekklesia for the election of generals adopted a citizenship decree. By contrast, in the case of synkletoi ekklesiai, a word or phrase indicating a statutory meeting is lacking, except in the one restored by Tracy.

The most informative passage on the synkletos ekklesia occurs in Aischines 2 *Embassy* 70 ff., where he refers in general terms to synkletoi ekklesias (plural and obviously numerous) as more in number than τὰς τεταγμένας ἐκ τῶν νόμων (2.27) in the wars with Philip and the wars in the Thermaic Gulf, starting with Amphipolis. He explains the conditions: σφαλερὰ καὶ ἐπικίνδυνα τὰ πράγματα. The period covered is about eleven years.[31] The Aris-

---

ἐκκλησία, and the content of the speech is clearly not directed at the military alone. In his commentary on 2.59.3, Hornblower discusses and favors Mogens H. Hansen's attempt to draw a distinction between a *military* meeting and an ordinary one. This seems to me completely false. See P. J. Rhodes, *The Athenian Boule* (Oxford 1972) 43–46, for a discussion of the relationship between the strategoi and the boule. Plutarch (*Mor.* 813d) says that the στρατηγία at Athens was the most important magistracy, comparable to the prytany in Rhodes and the boiotarchia in Boiotia.

[30]Meritt, *Klio* 52 (1970) 277–282, has shown that the phrase ἐκκλησία ἀρχαιρεσία κατὰ τὴν μαντείαν τοῦ θεοῦ could fall in any prytany after the sixth depending on some εὐσημία (Aristotle *Ath. Pol.* 44.4) which, he argues, was other than weather phenomena. Cf. C. Habicht, *Hesperia* 59 (1990) 568 n. 30, and P. J. Rhodes, *Commentary on the Aristotelian A.P.* (1981) 536–537. Ekklesias in progress might be abruptly adjourned, if unfavorable heavenly signs appeared: Thucydides 5.45.4–46.1 ("But before anything was ratified an earthquake occurred, and this ekklesia was adjourned. In the ekklesia on the next day ..."); Plutarch *Nikias* 10.6 ("Luckily for Nikias an earthquake intervened and the ekklesia was dissolved. On the following day ..."). Such ekklesias were lampooned by Aristophanes *Ach.* 170–171; *Clouds* 579–580; *Eccl.* 742.

[31]Cf. W. W. Goodwin, *Demosthenes on the Crown* (1901) 231. Hansen, throughout his numerous studies, makes the most inordinate error of taking the Aischines passage out of context in his argument that the synkletos ekklesia was not an extra meeting, and assumes that it applies exclusively

totelian rule applied to normal times. It is a mistake to assume that no exceptions were made in dangerous times.

To conclude this survey of ekklesias called synkletos, the passages in the text of Demosthenes (19.123) and Aischines (2.72) prove that there were extraordinary ekklesias called synkleteos. To argue that if Derkylos had arrived in Athens, not on Skirophorion (XII) 27, a date which suggests that this was the last regular meeting of the ekklesia, but on the 28th, the demos would have waited until some day in the next prytany (and year) to accommodate some rule that there could be only four ekklesias per prytany, is absurd.

I add that an obvious candidate for a synkletos ekklesia is described by Demosthenes (18 *Crown* 169), where unforeseen events necessitated the immediate calling of an ekklesia (Loeb tr.):

> Evening had already fallen when a messenger arrived bringing to the presiding councillors the news that Elatea had been taken. They were sitting at supper, but they instantly rose from table, cleared the booths in the market-place of their occupants, and unfolded the hurdles, while others summoned the commanders and ordered the attendance of the trumpeter. The commotion spread through the whole city. At daybreak on the morrow the presidents summoned the Council to the Council house, and the citizens flocked to the place of assembly.

Apparently, the trumpeter was to circulate and announce the special meeting throughout Attika. Diodoros (16.84.3) writes of the same event (Loeb): "Taken aback by the unexpected development, the Athenian generals summoned the trumpeters and ordered them to keep blowing the alarm signal the whole night through." The Loeb editor reads τὰ γέρρα ἀνεπετάννυσαν, but

---

to 347/6, the year of the archonship of Themistokles. See, for example, his *Athenian Assembly in the Age of Demosthenes* (Oxford 1987) 28–30.

all MSS. have ἐνεπίμπρασαν, a reading which has given rise to a large bibliography summarized by H. Wankel, Vol. 2 pp. 849–854.

* * * * * * * *

No student of calendar equations has endorsed a theory that the ekklesia met according to days in the festival calendar. Tables I, II, and IV in Chapter I present the difficulties any civic body would have in all periods, particularly in the Hellenistic. The theory seems to be based on statements in rhetoricians of the third and fourth centuries of our era. In the study of the comparative use of festival and prytany calendars on pp. 22–24 above, the suggestion was offered that in a state which set aside so many days to some form of religious devotion citizens instinctively came to refer to a calendar regulating these acts. I doubt that any modern scholar would have supported the theory found primarily in the scholia to Demosthenes and Aischines if the two orators had not given dates in the festival calendar. The theory has been buttressed by the claim that counting was inclusive in Demosthenes 24, a claim that I have attempted to refute.

INDICES

I. INDEX OF INSCRIPTIONS

II. INDEX OF ANCIENT AUTHORS

III. INDEX OF SELECTED SUBJECTS